Advanced
Nanodielectrics

Advanced Nanodielectrics
Fundamentals and Applications

edited by
Toshikatsu Tanaka and Takahiro Imai

Investigating R&D Committee
on Advanced Polymer Nanocomposite Dielectrics

The Institute of Electrical Engineers of Japan

Published by

Pan Stanford Publishing Pte. Ltd.
Penthouse Level, Suntec Tower 3
8 Temasek Boulevard
Singapore 038988

Email: editorial@panstanford.com
Web: www.panstanford.com

British Library Cataloguing-in-Publication Data
A catalogue record for this book is available from the British Library.

Advanced Nanodielectrics: Fundamentals and Applications

Copyright © 2017 Pan Stanford Publishing Pte. Ltd.

All rights reserved. This book, or parts thereof, may not be reproduced in any form or by any means, electronic or mechanical, including photocopying, recording or any information storage and retrieval system now known or to be invented, without written permission from the publisher.

For photocopying of material in this volume, please pay a copying fee through the Copyright Clearance Center, Inc., 222 Rosewood Drive, Danvers, MA 01923, USA. In this case permission to photocopy is not required from the publisher.

ISBN 978-981-4745-02-4 (Hardcover)
ISBN 978-1-315-23074-0 (eBook)

Printed in the USA

Contents

Preface   xix

1. Introduction: Attractiveness of Polymer Nanocomposites   **1**

Toshikatsu Tanaka

1.1   Nanocomposites Are Produced by Addition of a Small Amount of Fillers   2

    1.1.1   Methods of Mixing Quasi-Cubic Nanofillers   3

       1.1.1.1   Mixing methods   3

       1.1.1.2   Sol-gel method   3

    1.1.2   Methods of Mixing Layered Nanofillers   4

       1.1.2.1   Mixing of layered nanofillers by the intercalation method   4

       1.1.2.2   Mixing of layered nanofillers without an organically modifying process   4

1.2   Wide Applications Are Expected   4

1.3   Dielectric and Insulating Properties Exhibit a Drastic Change with Nanofiller Addition   5

1.4   A Balance Is Needed with Other Engineering Properties   9

    1.4.1   Thermal Expansion Coefficient and Withstanding Electric Stress   9

    1.4.2   Thermal Conductivity and Withstanding Electric Stress   10

1.5   Interfaces Dominate Bulk Properties   10

1.6   Nanocomposites Are a Derivative from Nanotechnology and Colloid Science   12

    1.6.1   Concept of Nanotechnology   12

    1.6.2   Interfaces Clarified by Colloid Science   13

	1.6.3	Technology Transferred from Composite Materials	13
	1.6.4	Birth of Polymer Nanocomposites	14
1.7	Bright Future of Nanocomposite World		14

PART 1: APPLICATIONS

2. Potential Applications in Electric Power and Electronics Sectors — 19

Takahiro Imai, Yoshiyuki Inoue, Masayoshi Nagata, Tetsuo Yoshimitsu, Kazutoshi Goto, Takanori Kondo and Takashi Ohta

2.1	Power Apparatus and Cables		19
	2.1.1	SF_6 Gas Has Excellent Performance but Also Has Greenhouse Effect	20
	2.1.2	Environmentally Friendly Apparatuses Are Developed without Use of Greenhouse Effect Gases	23
	2.1.3	Nanofiller Dispersed Insulating Materials Are Being Developed for Power Apparatuses	25
		2.1.3.1 Development of nanomicrocomposite for solid insulated switchgear	25
		2.1.3.2 Development of combined insulation system for molded instrument transformer	26
		2.1.3.3 Development of high-thermal-conductivity insulation material for all solid insulated substation	28
	2.1.4	DC Is Better Than AC for Long-Distance Power Transmission	29
	2.1.5	Space Charge Accumulation Is a Big Nuisance for DC Transmission Cables	31
	2.1.6	Nanocomposite XLPE Insulated Power Cables Have Been Developed	31

2.2	Magnet Wires and Motor Windings		38
	2.2.1	Breakdown of Motor Insulation Due to Partial Discharge by Inverter Surge	38
	2.2.2	Partial Discharge Inception Voltage by Inverter Surge Determined by Various Combined Factors	41
	2.2.3	Electric Characteristics of Nanocomposite Wire that Significantly Depend on Dispersion State of Nanofiller	45
	2.2.4	Dramatic Improvement of Inverter Surge-Resistant Life by Nanocomposite	47
	2.2.5	Mechanism by Which Nanofiller Suppresses Partial Discharge Degradation of Enameled Wire	49
	2.2.6	Application of Nanocomposite Surge-Resistant Wire to Production-Model Motors and Movement to International Standardization	50
2.3	Polymer Insulators for Outdoor Use		54
	2.3.1	Polymer Insulators Are Lightweight, Composite Structures	54
	2.3.2	What Kinds of Performances Are Needed for Polymer Insulators?	55
	2.3.3	Erosion Resistance Is Improved by Using Nanocomposites	57
	2.3.4	Bonding in Interfaces Is Strengthened by Nanofiller Addition	59
	2.3.5	Technological Evolution Is Expected for Nanocomposite Insulators	63
2.4	High-Density Mounted Components for Electronic Devices		68
	2.4.1	Polymer Insulators Are of Light-Weight and Composite Structures	68
	2.4.2	Validity of Nanocomposites as a Sealing Resin of Electronic Components	70
		2.4.2.1 Sealing resins for semiconductor devices and packaging	70

| | | 2.4.2.2 | Die bonding materials for semiconductor flip chips and underfill materials used to fabricate in their mounting semiconductor flip chips | 75 |

| | | 2.4.2.3 | High-dielectric-constant capacitor materials | 78 |

| | 2.4.3 | High-Heat-Dissipation, High-Thermal-Conductivity Nano-Microcomposites as Insulating Substrates for Electronic Equipment | | 79 |

| | | 2.4.3.1 | High-heat-dissipation, high-thermal-conductivity insulating substrates | 79 |

| | | 2.4.3.2 | Solder resist films for a semiconductor package for solder resist films | 83 |

3. Compatibility of Dielectric Properties with Other Engineering Performances 87

Toshikatsu Tanaka, Toshio Shimizu, Muneaki Kurimoto, Yoshimichi Ohki, Norio Kurokawa and Kenji Okamoto

3.1 Composite Materials with Both Ample Thermal Conductivity and Voltage Endurance 87

    3.1.1 Thermal Conductivity Measured by Laser Flush Method 88

    3.1.2 Thermal Conductivity Increased by Loading Micro-Filler into Polymers 89

    3.1.3 Further Increase in Thermal Conductivity Is Realized by Improvement of Interfaces 92

    3.1.4 Voltage Endurance Decreased by Micro-Filler Loading 94

    3.1.5 Recipe for Compatibility: An Exquisite Combination of Nano- and Microfillers 95

    3.1.6 Composite with High Thermal Conductivity and Sufficient Voltage Endurance 98

3.2 Composites with Both Low Thermal Expansion Coefficient and High Voltage Endurance 100

		3.2.1	Thermal Expansion Coefficient Is an Important Material Property of Molded Products	100

3.2.1 Thermal Expansion Coefficient Is an Important Material Property of Molded Products 100

3.2.2 It Is Possible to Decrease the Thermal Expansion Coefficient and to Improve the Electric Stress Resistance Together by Applying Nanocomposites 102

3.2.3 Hybrid Filling with Nano- and Microfillers Realizes Further Low Thermal Expansion and Drastically Improves Electrical Stress Resistance 105

3.3 Composites with High Permeability and High Dielectric Permittivity 110

3.3.1 For What Are Magnetized Dielectrics Used? 110

3.3.2 What Kind of Magnetized Dielectrics Are Available? 111

3.3.3 Let's Look at an Example under Research 112

3.3.3.1 For what purpose? 112

3.3.3.2 How to make the nanocomposite? 112

3.3.3.3 Let's look at important properties 113

3.4 High-Heat-Resistant Composites 118

3.4.1 Research Is Progressing in High-Heat-Resistant Composites by Using Nanocomposites 118

3.4.1.1 Polyethylene 118

3.4.1.2 Silicone resin 119

3.4.1.3 Epoxy resin 120

3.4.2 Thermal Endurance Varies with Methods of Nanofiller Dispersion 120

3.4.3 Practical Implementation of High-Thermal-Endurance Composites Is Now in Progress 121

3.5 Composites with High or Low Permittivity 129

3.5.1 Are Composites with High or Low Permittivity? 129

| | | 3.5.1.1 | High-dielectric-constant materials | 129 |

3.5.1.1 High-dielectric-constant materials 129

3.5.1.2 Low-dielectric-constant materials 131

3.5.2 Permittivity Can Be Increased by Dispersion of High-Dielectric-Constant Nanofiller? 132

3.5.3 Permittivity Can Be Lowered by Low-Permittivity Nanofiller? 135

PART 2: FUNDAMENTALS (MATERIAL PREPARATION)

4. Preparation of Polymer Nanocomposites: Key for Homogeneous Dispersion **141**

Mikimasa Iwata, Yuki Honda, Takahiro Imai and Minoru Okashita

4.1 Reactive Precipitation Method: Sol-Gel Method 141

4.1.1 The Sol-Gel Method Can Result in Excellent Dispersion of Nanofillers in Polymers 142

4.1.2 Good Recipe and Special Care Are Needed for the Sol-Gel Method 143

4.1.3 What Kind of Mechanisms Are Working to Obtain Various Characteristics? 145

4.1.4 Composites Produced by the Sol-Gel Method Are Used in Daily Life 147

4.2 Mixing Methods for Quasi-Spherical Fillers 150

4.2.1 Quasi-Spherical Nanofillers to Be Used Are Extremely Fine 150

4.2.2 Various Resins Are Used for Polymer Nanocomposites 151

4.2.3 Polymer Nanocomposites Can Be Derived by Dispersing a Quasi-Spherical Nanofiller into a Polymer 151

4.2.3.1 Dispersing a quasi-spherical nanofiller in a thermosetting resin 151

4.2.3.2 Dispersing a quasi-spherical nanofiller in a thermoplastic resin 153

		4.2.3.3	Making a polymer nanocomposite from colloidal silica	153
		4.2.3.4	Nanoparticles can be dispersed by ultrasonic waves and centrifugal force	154
	4.2.4	Control of Nanofiller Diameter Size Is Important for Good Nanocomposites		155
4.3	Reactive Mixing Method for Fillers with Layered Structures			158
	4.3.1	Unit Layer of Fillers with Layer-Structured Fillers Is 1 nm in Thickness		158
	4.3.2	Organic Compounds Can Be Brought in between Neighboring Layers		159
	4.3.3	Layer-Structured Fillers Are Exfoliated and Dispersed		162
	4.3.4	State of Filler Dispersion Is Affected by Various Factors		163
	4.3.5	Various Techniques Are Developed for Homogeneous Dispersion		166
		4.3.5.1	Dispersion of clays swelled with organic solvent	166
		4.3.5.2	Clay dispersion using AC voltage application	167
		4.3.5.3	Clay dispersion using high-shear compounding in solid phase	168
		4.3.5.4	Dispersion of clays without organic modification	169
4.4	Surface Modification of Nanofiller Helps Uniform Dispersion			171
	4.4.1	Surface Modification Is Important		171
	4.4.2	Several Methods Are Available for Surface Modification		173
	4.4.3	Larger Filler Particles May Be Surface-Modified by Nanofillers		177

PART 3: FUNDAMENTALS (MATERIAL CHARACTERISTICS)

5. Drastic Improvement of Dielectric Performances by Nanocomposite Technology   **183**

Muneaki Kurimoto, Kazuyuki Tohyama, Yasuhiro Tanaka, Yoshinobu Mizutani, Toshikatsu Tanaka, Masayuki Nagao, Naoki Hayakawa, Takanori Kondo and Tsukasa Ohta

5.1   Permittivity and Dielectric Loss: Dielectric Spectroscopy   183

     5.1.1   Permittivity and Dielectric Loss Are Evaluated by Dielectric Spectroscopy Dependent on Frequency and Temperature   184

     5.1.2   Permittivity of Microcomposites Is Determined by Composition Ratio   188

     5.1.3   Permittivity Rises or Falls Due to Nanofiller Inclusion   190

     5.1.4   Much Attention Is Paid to an Unexpected Strange Phenomenon of Permittivity Decrease   194

     5.1.5   How Do Nanofillers Act for Permittivity of Nanocomposites?   196

5.2   Low Electric Field Conduction   199

     5.2.1   Electrical Conductivity Is One of the Most Important Factors in Electrical Insulation   199

     5.2.2   Electrical Conductivity Increased by Nanofiller Addition in Some Cases   202

     5.2.3   Electrical Conductivity Decreases with Nanofiller Addition in Some Cases   204

5.3   Conduction Current under High Electric Field and Space Charge Accumulation   210

     5.3.1   Is Breakdown Unpredictable?   210

     5.3.2   Is Space Charge Accumulation Presage of Degradation or an Electrical Breakdown?   213

     5.3.3   Adding Nano-Size Filler to Significantly Reduce Space Charge Accumulation Even under High DC Stress!   217

     5.3.4   Why Does Addition of Nano-Size Filler Suppress Injection of Packet-Like Charge?   220

5.4		Short-Term Breakdown Characteristics	224
	5.4.1	Method of Measuring Short-Term Breakdown Characteristics	224
	5.4.2	How the Short-Term Breakdown Characteristics of Nanocomposite Insulators Change	226
	5.4.3	Filler State Is Important in Order to Improve Short-Term Breakdown Characteristics	228
5.5		Long-Term Dielectric Breakdown	233
	5.5.1	Treeing Breakdown of Polymers Is Evaluated in Terms of Tree Shapes and V–t Characteristics	233
	5.5.2	Treeing Breakdown Lifetime Is Enormously Prolonged by Nanofiller Addition	236
	5.5.3	What Role Do Nanofillers Play in Tree Initiation?	238
	5.5.4	A Crossover Phenomenon Appears: Tree Growth vs. Voltage	240
	5.5.5	How Do Nanofillers Act for Tree Growing Processes?	241
5.6		Insulation Degradation	244
	5.6.1	PD Resistance of Polymers Valuated in Terms of Erosion Phenomena	244
	5.6.2	PD Resistance of Polymers Is Drastically Increased Due to Nanofiller Inclusion	247
	5.6.3	What Are Mechanisms for PD Erosion in Nanocomposites?	249
5.7		Insulation Degradation	253
	5.7.1	Water Trees Are Generated When Polymers Are Subjected to Both Water and Electric Field	253
	5.7.2	Progress of Water Trees Is Suppressed by Nanofiller Addition	256
	5.7.3	What Role Do Nanofillers Play to Suppress Water Tree Progress?	258
5.8		Insulation Degradation	263

xiv | Contents

5.8.1 Tracking Takes Place by Surface Pollution of Insulators — 263

5.8.2 Inclined Plane Test and Arcing Test Are the Standard Test Methods to Evaluate Tracking and Erosion Resistance of Insulators — 264

5.8.3 Tracking and Erosion Resistance Is Greatly Enhanced by Nanofiller Addition — 265

5.8.4 Improvement of Heat Resistance Will Lead to That of Tracking and Erosion Resistance — 268

5.9 Material Degradation Due to Electrochemical Migration — 272

5.9.1 Why Is a Measure Against the Migration Necessary? — 272

5.9.2 What Kind of Phenomenon Is the Migration? — 273

5.9.3 Various Reliability Test Methods Are Available for the Migration — 275

5.9.4 The Migration Can Be Evaluated by Measuring the Distribution of Space Charge — 278

5.9.5 Nanocomposites Will Suppress the Migration — 280

6. Thermal and Mechanical Performance of Nanocomposite Insulating Materials — **285**

Satoru Hishikawa, Shigenori Okada, Takahiro Imai and Toshio Shimizu

6.1 Thermal Performance — 285

6.1.1 Thermal Characteristics Comprise Thermal Behavior, Thermal Properties, and Heat Resistance — 286

6.1.2 Epoxy Resin Thermal Characteristics Can Be Altered Using a Nanofiller — 287

6.1.2.1 Effects of several types of nanofillers on the glass transition temperature — 287

| | | 6.1.2.2 | Effects of several types of nanofillers on the glass transition temperature | 288 |

6.1.2.2 Effects of several types of nanofillers on the glass transition temperature — 288

6.1.2.3 Effects of nanofiller dispersion on the glass transition temperature of nanomicrocomposites — 290

6.1.3 Thermal Characteristics of Several Types of Polymers Change by the Dispersion of the Nanofillers — 292

    6.1.3.1 Room Temperature Vulcanizing of RTV silicones — 292

    6.1.3.2 Polypropylene — 293

    6.1.3.3 Polyethylene — 294

6.1.4 Polymer Thermal Characteristics Can Be Changed by Interfacial Interactions between Nanofillers and Polymer — 295

6.2 Mechanical Performance — 298

6.2.1 Polymer Composites with Improved Mechanical Performance Are Utilized in Daily Life — 298

6.2.2 Various Mechanical Properties Are Classified by Stressing Time Conditions — 300

    6.2.2.1 Force applied slowly at a constant speed — 301

    6.2.2.2 Force applied as an impact — 301

    6.2.2.3 Constant force applied for a long duration — 302

    6.2.2.4 Force applied periodically — 302

6.2.3 Tensile Strength Increases by Nanofiller Addition — 302

6.2.4 Flexure Performance Improves by Nanofiller Addition — 305

6.2.5 Nanofillers Will Suppress Propagation of Crack — 306

6.2.6 Various Mechanical Properties Are Reported for Dielectric Nanocomposites — 308

| | | 6.2.6.1 | Scratch hardness of polyamide nanocomposites | 309 |

6.2.6.1 Scratch hardness of polyamide
nanocomposites 309

6.2.6.2 Charpy impact strength of
epoxy resin/basket-shaped silica
nanocomposites 309

6.2.6.3 Mechanical characteristics of
nano- and microcomposites 310

6.3 Long-Term Characteristics 313

6.3.1 Thermal Endurance of Polymers Is
Improved by Nanocomposites 313

6.3.2 Fatigue Properties of Polymer Are
Improved by Nanocomposite 315

PART 4: THEORETICAL ASPECTS

7. Structures of Polymer/Nanofiller Interfaces 323

Toshikatsu Tanaka and Muneaki Kurimoto

7.1 Interfaces Have Volume 323

7.1.1 What Are Interfaces? 324

7.1.2 Interfaces Formed by Inorganic Fillers
and Organic Polymers: What Features
Do They Have? 324

7.1.2.1 Interfaces account for enormous
rate in nanocomposites 324

7.1.2.2 Silane couplings fuse inorganic
and organic matters into polymer
nanocomposites 325

7.1.2.3 What are interfacial
interactions like? 326

7.1.3 Various Interfacial Models Are Proposed 327

7.1.3.1 A simple two-layered interfacial
model 327

7.1.3.2 Multi-core model analogously
derived from quantum mechanics
and collide chemistry 328

7.1.3.3 Water shell model devised for
interfaces with minute gaps 330

	7.2	Physicochemical Analysis Methods for Interfaces	333
		7.2.1 Shapes, Sizes and Dispersion Are Evaluated by SEM and TEM Observation	333
		7.2.2 Filler Content Is Estimated by Measuring the Density of Nanocomposites	336
		7.2.3 Inter-Filler Distances Are Evaluated in Mesoscopic and Macroscopic Scales	337
		7.2.4 Some Methods Are Proposed to Investigate Organic and Inorganic Bonding States	341

8. Computer Simulation Methods to Visualize Nanofillers in Polymers: Toward Clarification of Mechanisms to Improve Performance of Nanocomposites — **345**

Masahiro Kozako and Atsushi Otake

	8.1	Non-Empirical Molecular Orbital Method	345
		8.1.1 Computer Simulation of Nanocomposites Has Begun	346
		8.1.2 What Is ab initio Calculation Like?	347
		8.1.3 What Can Be Found When Using the ab initio Calculation?	348
		8.1.3.1 Interfaces account for enormous rate in nanocomposites	348
		8.1.3.2 Electron trap sites can be estimated	349
		8.1.4 Application to Large-Scale Systems Is Also Attempted	351
	8.2	Simulation of Performance of Nanocomposites by Coarse-Grained Molecular Dynamics	352
		8.2.1 What Is Molecular Dynamics?	352
		8.2.1.1 Overall concept of molecular dynamics	352
		8.2.1.2 Advantages of coarse-grained molecular dynamics	354
		8.2.1.3 Calculation method	356
		8.2.1.4 Accurate determination of force fields for large analytical systems	358

		8.2.2	Examples of Practical Use of Coarse-Grained Molecular Dynamics	359
		8.2.2.1	How is a nanofiller dispersed in resin?	359
		8.2.2.2	Type of nanofiller and mechanical properties of resin	361
		8.2.2.3	What occurs in the vicinity of a nanofiller?	362

9. Epilogue: Environmental Concerns and Future Prospects   **365**

Takahiro Imai

9.1	Awareness in Dealing with Nanofillers		366
	9.1.1	Effects of Nanofillers on Human Body and Environment Are a Matter of Concern	366
	9.1.2	Risk Evaluation of Nanofillers Is in Progress	367
	9.1.3	Guidelines for Dealing with Nanofillers	369
9.2	Future Prospects		371
	9.2.1	Worldwide Interest Continues to Be Enhanced Year by Year	371
	9.2.2	Various Applications Are Explored for Their Realization	373
	9.2.3	Open the Door to Polymer Nanocomposites for Future	376
Index			381

Preface

It is well known that electricity etymologically originated from amber in Greece. Ancient Greeks knew about fricative electrostatic force created with amber. Historically, amber is considered to be one of the first insulation materials. Although primary electrical phenomena have been researched since time began, the invention of the Volta cell in 1800 accelerated the scientific study of electricity, which began in the early 1800s, and industrial applications of electricity have expanded since the late 1800s. Insulation technology has developed along with electrical technology. Natural materials such as cotton, rosin, and ceramic were used as insulation materials before the appearance of phenol resin in 1907, and synthetic polymers had an impact on the progress of insulation technology.

Composite technology combining polymers with other materials has been an effective approach to improving dielectric and insulation materials. The innovative concept of nanodielectrics appeared in the late 1900s. Many nanodielectrics studies dealing with the dielectric and insulation characteristics of polymer nanocomposite materials composed of polymers and inorganic particles with nanometric dimensions have mushroomed since the early 2000s. The National Nanotechnology Initiative started by the US President in 2000 also had a strong influence on the development of nanodielectrics.

Today, "nano" is a buzzword and has fascinated many researchers and engineers in various fields. Is "nanodielectrics" also a buzzword in the field of dielectric and insulation materials? Continuing development of nanodielectrics technology has produced more than 1,000 technical papers, and a lot of information has been accumulated. In fact, many papers have demonstrated that nanodielectrics offers many advantages and needs in-depth analyses to clarify a mechanism for the enhancement of dielectric and insulation properties. Moreover, applications of nanodielectrics are increasing rapidly, for example, in DC power cables for submarine transmission and magnet wires for motors. Who deems nanodielectrics to be fashionable in the dielectrics and electrical insulation field at present? The editors would like to provide readers with an opportunity to open the door to the new materials of "advanced nanodielectrics."

The Investigating R&D Committee of Advanced Polymer Nanocomposite Dielectrics of the Institute of Electrical Engineers of Japan (IEEJ) was established in 2003. The editors believe that the committee played a pioneering role in the investigation of nanodielectrics research. Investigations of the fundamental properties of dielectrics and their industrial applications have been conducted in the past decade. The committee published a Japanese book as a basis for investigation activities. This book was the winner of the IEEJ Outstanding Technical Report Award in 2016. The book entitled *Advanced Nanodielectrics: Fundamentals and Applications* is the translated version of the award-winning book.

The book provides detailed coverage on processing and the electrical, thermal, and mechanical properties of polymer nanocomposite dielectrics. Special consideration of the surface modification of inorganic particles with nanometric dimensions and theoretical aspects of interface and computer simulation promotes an understanding of the characteristics of nanocomposites. In particular, a large interface between the polymer and particles dominates properties in nanocomposite dielectrics, and that is illustrated in detail. Moreover, potential applications in the electric power and electronics sectors demonstrate the many advantages of nanocomposites. The book is a definitive and practical handbook for beginners as well as experts. The key features of the book are as follows:

- Provides knowledge about cutting-edge technologies in polymer nanocomposites to be used in electrical and electronics engineering
- Explains significant roles of "guest" nano-fillers in "host" polymer matrices
- Explains computer simulations in organic and inorganic composite materials
- Includes contributions from leading international academic and engineering experts

- Strikes a good balance between fundamental production methods and applications
- Includes many practical examples of polymer nanocomposites with many experimental results

Finally, the editors would like to express their gratitude for the cooperation of the people involved in this translation project. All contributors devoted considerable effort to translate the original Japanese book into English. Shin-ichi Yamazaki and the Technical Committee on Dielectrics and Electrical Insulation of the Institute of Electrical Engineers of Japan significantly supported this activity. The editors especially appreciate the assistance of Stanford Chong and Arvind Kanswal of Pan Stanford Publishing.

Toshikatsu Tanaka
Takahiro Imai
Editors

Chapter 1

Introduction: Attractiveness of Polymer Nanocomposites

Toshikatsu Tanaka

IPS Research Center, Waseda University,
Kitakyushu-shi, Fukuoka 808-0135, Japan

t-tanaka@waseda.jp

Polymer nanocomposites have 50 years of history. In the past, colloids containing nanometer-scale particles were studied as a related science, and nanotechnology has emerged in the recent past as a science to treat nanoscale substances. Polymer nanocomposites emerged from nanotechnology and have been intensively investigated in colloid science. Polymer nanocomposites were developed as engineering plastics in the 1990s and appeared as dielectric and insulating materials in the 2000s. It was found that various characteristics can be improved when a few inorganic nanofillers are added to pure polymer resins. In the dielectrics field, studies started from the latter half of the 1990s, and basic research and application development have been carried out in the 2000s. Thanks to this R&D trend, various kinds of polymer nanocomposites were investigated to determine their applicability as electrical insulation for electric power equipment, power

---

Advanced Nanodielectrics: Fundamentals and Applications
Edited by Toshikatsu Tanaka and Takahiro Imai
Copyright © 2017 Pan Stanford Publishing Pte. Ltd.
ISBN 978-981-4745-02-4 (Hardcover), 978-1-315-23074-0 (eBook)
www.panstanford.com

cables, and microelectronic devices. Characteristics of these super-composites with multiple functions and high performance should be explored in response to social needs.

1.1 Nanocomposites Are Produced by Addition of a Small Amount of Fillers

Nanocomposites are generally composed of host materials and guest materials. This book deals with the combination of a polymer as a host with an inorganic filler as a guest. They are defined so that they may include fillers of less than 100 nm in at least one dimension. The amount of added fillers is generally less than 10 wt% (weight %).

Nanocomposites are created to provide guest materials with excellent characteristics while retaining original good characteristics of the host materials. With this nanocomposite technology, it might be possible to create a super-material even with both hard and elastic properties, which are usually contradictory. Generally, inorganic materials have excellent optical, electrical, mechanical, and thermal properties, while organic materials have other excellent properties such as lightness, flexibility, and moldability. Composite materials are now expected to have mutually complementary characteristics and novel properties that neither host nor guest materials possess on their own.

Three combinations of the host and the guest are available: inorganic with inorganic, inorganic with organic, and organic with organic. One of the most attractive combinations is considered to be an organic substance (polymer) as a host with an inorganic substance as a guest. The success of organic/clay composites opens a nanomaterial world. In the beginning, mechanical and optical properties attracted much attention and gradually interest shifted with the progress of technology toward other important characteristics such as gas barrier performance, lubrication ability, heat resistance, heat dissipation, electrical conductivity, and insulating properties.

Nanocomposites need nanofillers. There are three fabrication methods: a solid-phase method, a liquid-phase method and, a vapor-phase method. The conventional method for nanofillers is the solid-phase method, which is enabled by crushing bulk materials. This method is not suitable for the manufacture of multi-component-type fillers with multi-chemical elements and is limited to particle size down to 0.1 μm (100 nm). The liquid-phase method and the vapor-phase method are better for synthesizing nanofillers. Incidentally, the main manufacturers for nanofillers are Aerosil, C. I. Kasei, and the like, which manufacture nanofillers of SiO_2, Al_2O_3, and TiO_2. No production techniques are available in the public domain.

The next issue is related to an interesting topic of how to mix nanofillers into polymers. There are two kinds of nanofillers: quasi-cubic and layer-structured. First, there are two main methods for mixing quasi-cubic nanofillers into polymers: the direct mixing method and the sol-gel method. Second, methods of mixing layered nanofillers are accompanied by chemical reactions.

1.1.1 Methods of Mixing Quasi-Cubic Nanofillers

1.1.1.1 Mixing methods

(i) Epoxy nanocomposites

Three kinds of materials, an epoxy resin (main agent), nanofillers, and silane coupling agents, are first mixed by using a high-pressure homogenizer that has the ability to produce high shearing power, and then a hardener is added. The resultant mixture is cast, vacuum evacuated, and heat cured.

(ii) Polyethylene nanocomposites (made by one of the available methods)

Nanofillers are put into chloroform (a solvent) and dispersed under ultrasonic irradiation. This dispersion mixture is added into a xylene solution of low-density polyethylene (LDPE). After this solution is stirred for 2 h, the solvent is volatized under decompression (100 Pa, 393 K) to finally produce a sample material.

1.1.1.2 Sol-gel method

The sol-gel method is a method to synthesize and disperse fine glassy particles chemically in a polymer by reactions in the sol-gel process, i.e., the hydrolysis dehydration condensation reaction. Specifically, metal alkoxides (e.g., alkoxysilane) are hydrolyzed or condensed with an alkali-catalyst. Generally, tetraethoxysilane

(TEOS) or tetramethoxy-silane (TMOS) is used to precipitate SiO_2 or nano particles in a polymer.

1.1.2 Methods of Mixing Layered Nanofillers

Layered nanofillers (layered silicates or clays) are composed of repeated layers of a basic unit layer structure consisting of three sheets: a Si-O tetrahedral sheet, an Al-O/Mg-O octahedron sheet, and a Si-O tetrahedral sheet. Their unit silicate layer is ca. 100 nm in length, ca. 10 nm in width, and ca. 1 nm in thickness. Negative charge is generated by substituting silicon in the Si-O tetrahedral sheet with aluminum or trivalent cations in the central octahedral sheet with divalent cations. In order to compensate for this negative charge, cations such as Na^+ are bonded to interlayers. By this structure, layered nanofillers possess ion exchange ability and can therefore form intercalation compounds with organic substances.

1.1.2.1 Mixing of layered nanofillers by the intercalation method

Layered nanofillers (layered silicate) are mixed into polymers by the intercalation method. This can be mainly divided into two processes: an interlayer insertion process and a layer exfoliation process.

1.1.2.2 Mixing of layered nanofillers without an organically modifying process

Layered nanofillers in water slurry are compounded with polyamide in a biaxial mixing extruder to disperse layered nanofillers.

1.2 Wide Applications Are Expected

DC conductivity, permittivity, dielectric loss, high electric field conductivity, formed space charge, short time dielectric breakdown strength (DC, AC and Impulse), long-term breakdown characteristics (treeing lifetime, partial discharge resistance, tracking-resistance), and the like have been investigated for dielectric materials obtained through nanocomposite fabrication technology. As a result, various desirable results have been obtained and a number of researchers have participated in this field. From the engineering standpoint, much research has begun to develop power cables, motor winding insulation (magnet wires), rotating machine winding insulation, switchgear insulation, polymer insulators, and microelectronics-use printed wiring boards (PWB). Table 1.1 shows a possible wide range of applications of polymer nanocomposite insulation with various characteristics.

Table 1.1 Wide range of applications of polymer nanocomposites

Application	Host materials	Guest nanofillers	Improvement	Purposes
Rotating machine winding insulation	Epoxy	Nano alumina	PD resistance	Compact
Switchgear insulation	Epoxy	Layered silicate	Dielectric strength	Compact
Printed wiring boards	Epoxy	Nano silica	Dielectric strength, thermal conductivity, head resistance	Lower cost
Magnet wire insulation	Polyesterimide, polyamideimide	Nano silica	PD resistance	Against surge voltage
HV AC power cable	XLPE	Nano silica	Dielectric strength	Compact
HV DC power cable	XLPE	Nano magnesia	Space charge	DC characteristics
Capacitors	Polypropylene	Nano silica	Dielectric strength	Compact
Polymer insulators	Silicone rubber	Nano silica	Tracking resistance	Lower cost

1.3 Dielectric and Insulating Properties Exhibit a Drastic Change with Nanofiller Addition

Target organic materials for polymer nanocomposites are generally epoxy resins as general-purpose insulating materials and also include cable insulation polyethylene and XLPE (cross-linked polyethylene), capacitor dielectric polypropylene, and polymer insulator silicone resins. Inorganic nanofillers used for nanocomposites are silica (SiO_2), layered silicates, titania (TiO_2), alumina (Al_2O_3), boehmite alumina (AlOOH), silicon carbide (SiC), magnesia (MgO), etc. Nanocomposites are fabricated by certain methods such as the direct mixing method, the sol-gel method, and the intercalation method using organic and inorganic materials chosen for specific purposes. Silane coupling agents are generally used to chemically bind organic base polymers with inorganic fillers. Table 1.2 summarizes excellent characteristics of polymer nanocomposites that have been achieved to date.

Table 1.2 Excellent performance of polymer nanocomposites

Characteristics	Degree of improvement	Remarks
PD resistance	Enormous increase	Nano segmentation of resins
Treeing lifetime	Enormous extension	Especially under moderate electric stress
Space charge	Enormous reduction	Increase of threshold electric stress
Tracking resistance	Great increase	Similar to PD resistance
Dielectric breakdown	More or less increase	Residual cross-linking agents and impurities
Permittivity	Increase/decrease	Decrease under a certain condition
Dielectric loss	Temperature and frequency dependences	Interfacial structures and impurities
DC conductivity	Increase/decrease	Carrier traps and impurities

Thanks to such favorable qualitative results, fundamental research has been deepened and engineering application research has started.

It is indicated in Table 1.2 that substantial improvement is made for partial discharge resistance if polymer resins include a small amount of nanofillers. Since partial discharges take place not only in high-voltage power apparatuses but also in microelectronics devices, these improved characteristics are significantly important and beneficial to power and microelectronics sectors.

Treeing is a phenomenon in which tree-like dendritic paths are formed in electrical insulation when it is subjected to high electric field. It is considered to be an index for electric field resistance of insulating materials. Treeing resistance is usually evaluated using a needle-plane electrode system. A needle tip is arranged vertically to the plane of a counter flat electrode by keeping a certain distance between them, say 1 to 3 mm, in a solid dielectric. A tree starts from the tip of the needle electrode with a tip radius of 3 to 5 μm. V–t characteristics are obtained and evaluated until tree initiation or breakdown failure. When nanofillers are added into polymer base resins, generated trees change in shape and retard in initiation and propagation. As a result, time to long-term failure is lengthened and dielectric breakdown strength is enhanced. It was elucidated that treeing lifetime is extended up to 100 times under a design (moderate) electric field. Investigation was made of many kinds of composite systems such as epoxy/titania, epoxy/alumina, epoxy/layered silicate, epoxy/boehmite alumina, and polyethylene/magnesia.

Space charge is easily formed inside when a DC high electric field is impressed on solid insulation, resulting in a reduction of breakdown strength. The space charge should be much reduced in order to develop DC XLPE cable insulation as one of the engineering applications. Nondestructive measurement methods to measure space charge in a solid were developed about 20 years ago. They include the pulsed electro-acoustic (PEA) method, the laser-induced pressure pulse (LIPP) method, and the heat pulse method. The PEA method is most popular probably because it is easy to handle and is reliable.

Space charge is accumulated in a solid insulation when it is subjected to a high electric field. This space charge will distort the electric field distribution inside and create a higher electric field locally than the average. The insulating substance may fail electrically under a lower voltage than is usually expected. Space charge formation processes have been investigated in LDPE and its nanocomposite with magnesia (LDPE/MgO). It was clarified that this nanocomposite possesses higher breakdown strength than its base LDPE under high DC stress. Under a high electric field, a packet charge appears in LDPE but disappears in its nanocomposite. When a packet charge is generated, dielectric breakdown is likely to take place. Some mechanisms are proposed for this type of space charge behaviors. According to these concepts, charge injection is suppressed either by the added nanofillers or by the nanofiller-induced traps and dipoles.

Breakdown strength is one of the most important characteristics for insulation design. Insulation design is made by this performance for high-voltage equipment. Theoretically, the intrinsic breakdown strength of solid insulating materials is the range from several MV/cm to 10 MV/cm at room temperature. Practically, breakdown strength is far lower than this value and is dependent on insulation thickness, material performance (such as defects, voids, amorphous regions, molecular lengths, and impurities), types of voltage (such as AC, DC, and impulse), and ambient conditions. In addition, data scatter is relatively large, so under this situation breakdown strength data are often processed by Weibull statistics. It is reported that the breakdown strength is increased by adding nanofillers. Several reports have demonstrated the following positive results. AC breakdown strength increases by 25% when adding nanofillers to LDPE. Three kinds of nanofillers, sepiolite clay (SEP), montmorillonite clay (MMT), and their combinations, certainly improve $V-t$ characteristics. MMT nanofillers exhibit the highest breakdown strength of all. They are considered to contribute most to the increase of short time breakdown strength, since they have a high aspect ratio. SEP is considered to contribute to the homogeneous dispersion of nanofillers. Similar results were obtained for a system of XLPE/vinylsilane treated silica nanocomposite.

A different kind of nanofiller is used to improve DC and impulse voltage breakdown strength of LDPE. An example is MgO. Addition of this filler increases DC and impulse breakdown strengths by 32% and 19%, respectively. However, titania contributes negatively. Even in this case, if titania is dried and/ or processed by silane, breakdown strength reduction can be mitigated. On the other hand, titania can increase the breakdown strength of epoxy resins as compared to LDPE. Nanofiller addition to epoxy resins increases DC, AC, and impulse breakdown strength by 18%, 9%, and 2%, respectively, but micro-filler loading reduces breakdown strength by 10%, 16%, and 2%, respectively. However, it is reported that even epoxy nanocomposites (nano silica, nano alumina and nano zinc oxide) show reduced AC breakdown strengths in an inverse way. Optimum conditions should be sought to fabricate nanocomposites, because breakdown strengths are likely be influenced by various conditions such as filler size, drying conditions, interfacial modification, dispersing methods, dispersing conditions, and hardening conditions.

1.4 A Balance Is Needed with Other Engineering Properties (Creation of Nanomicrocomposites)

1.4.1 Thermal Expansion Coefficient and Withstanding Electric Stress

Metal conductors are molded with polymer resins into one component to produce power apparatuses such as switchgear with a function of interrupting electric current, where the thermal coefficient of the resins should be equalized to that of the metals as closely as possible. Since polymers like epoxy resins have larger thermal coefficients than metals like aluminum, the thermal coefficient of the resins is generally reduced and adjusted to that of the metals by filling the resins with inorganic micro-scale fillers (silica, etc.). Since polymer nanocomposites with excellent electric stress withstanding performance have similar thermal expansion coefficients to their base epoxy resins, they are not applicable just as they are. Methods of filling them with micro-silica fillers have been investigated in order to maintain original withstanding performance and at the same time reduce thermal expansion coefficient. Thus, micro-silica-filled nanocomposites with nano-silica or layered silicate are fabricated. This technology enables the development of insulation support parts made of nanomicrocomposites to make a whole switchgear system more compact. Moreover, epoxy resins are hard and fragile in nature, and the addition of nanofillers is also effective in improving such performance and increasing their tenacity. Epoxy resins are filled with nanofillers like organically modified clay (layered silicate) in order to increase their mechanical strength. This effectiveness depends on methods for clay processing and is the highest when they are mixed while they are swollen with polar solvent and stressed with shear force.

1.4.2 Thermal Conductivity and Withstanding Electric Stress

It is desirable that electronics packages and PWB made of polymer resins have high thermal conductivity. To make epoxy resins more thermally conductive, a method of infilling micro fillers such as alumina with high thermal conductivity is generally utilized. Infilling of a lot of micro-fillers is needed for that purpose, which is rather likely to reduce electric stress withstanding. Once lowered, electric stress withstanding recovers if a small amount of nanofillers is added. For example, epoxy resins can gain high thermal conductivity like 10 W/(m·K) by infilling BN (boron nitride) micro-fillers and can retain sound electric stress withstanding when they are further infilled with nano-silica.

1.5 Interfaces Dominate Bulk Properties

Polymer nanocomposites are compounds of organic polymers with inorganic nanofillers. They are characterized by enormously large interfaces between two kinds of immiscible substances. That is, the excellent performance that polymer nanocomposites exhibit is considered to originate from the interfaces between the two substances. For example, the surface density of the interfaces is as large as 3.5 km^2/m^3 when nanofillers 40 nm in diameter are infilled by 5 wt%. Interfaces have a certain thickness and this region is called an interaction zone. If this zone has 10% of the filler diameter, it occupies 50% of its total volume. This indicates that this volume, i.e., the formed interfaces, will dominate the performance of the nanocomposite materials.

Figure 1.1 shows a sketch of two possible interface structures and a core model. These interfaces are different from the original polymers in their morphology and are considered to make a tremendous contribution to the macroscopic characteristics of the resulting nanocomposites. The above three structures are explained below. Polymer chains are randomly distributed around the surface of a nanofiller in Fig. 1.1a. Polymer chains are coordinated more or less perpendicular to the surface of a nanofiller particle in Fig. 1.1b, which is similar to spherulite structures. There is another possible option in which interfaces consist of polymer chains layered in parallel around a filler particle. Figure 1.1c shows a three-core model consisting of multi-layers with morphologies in which bonding structures are different.

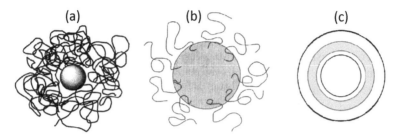

Figure 1.1 Interfacial structures between nanofillers and polymer matrices.

These interfacial structures are expanded in the radial direction and therefore form an interaction with some thickness. The thickness of such an interface is considered to consist of multiple layers with different properties, which is often explained in terms of some two-layer model. A multi-core model is also proposed, as indicated in Fig. 1.1c (a three core model in this case). These interfaces are expected to exhibit respective mesoscopic characteristics. For that purpose, it is necessary to clarify mesoscopic characteristics such as morphology by positively using a variety of excellent physic-chemical analysis methods.

Interactions are generally caused by forces working between polymers and fillers: ordinary chemical bonds, hydrogen bonds, van der Waals force, and even mechanically structural forces like anchor force. Therefore, such interactions in their appearance and strength are controlled by chemical and physical factors, which are chemical structures such as chemical bonds and polar radicals, and filler characteristics such as their surface areas or surface/volume ratio, surface roughness, chemical compositions and functional groups (radicals such as =O, –COOH, and –OH). It is also important to investigate how sound "wettability" is between inorganic fillers and organic polymer matrices, where it is actually represented by the total area of "wetting." Accordingly, methods of strengthening interactions are required to make the conditions of filler dispersion as optimum as possible and to improve the wettability of filler surfaces, which are the knowhow for compounding technologies. Surface treatment is effective for this purpose. The concept of a bound polymer is proposed to explain how mechanical interactions are created between a filler particle and its surrounding polymer matrix. This indicates that the matrix polymer chain is constrained by the surfaces of the fillers in their free movement, resulting in formation of an immobile layer, i.e., a bound polymer layer around the filler surface. As this layer is considered to become thicker as the interaction becomes stronger, the thickness represents a measure of the interaction, if it can be quantitatively measured. Values of 10 to 200 nm are obtained for a PVC (polyvinylchloride)–inorganic filler composite substance.

1.6 Nanocomposites Are a Derivative from Nanotechnology and Colloid Science

1.6.1 Concept of Nanotechnology

Nanocomposites are recognized as the successor to bottom-up nanotechnology. Richard P. Feynman, a Nobel Laureate of Physics, stated "There's plenty of room at the bottom" at the annual meeting of the American Physical Society held at California Institute of Technology in 1959 and proposed the concept of a new scientific technology about minute space. It is generally accepted that this is the origin of nanotechnology. This is mainly due to K. Eric Drexler's current activity. Then, Mr. Norio Taniguchi used the word "nanotechnology" for the first time at "the International Conference on Production Engineering (ICPE)," which was held in Japan in 1974. Furthermore, K. Eric Drexler proposed molecular nanotechnology in 1986 in the book *Engines of Creation: The Coming Era of Nanotechnology*. This was considered to be an ultimate method as bottom-up technology for constructing new materials from nanometer size up to macroscopic size. Later the "National Nanotechnology Initiative (NNI)" issued in the United States in the year 2000 influenced not only the United States but also many other countries to accelerate R&D in nanotechnology. In Japan, nanotechnology is ranked as a national high priority R&D

item by the Council for Science, Technology and Innovation of the Cabinet Office of Japan to be explored through industry–government–academia cooperation under the initiative and guidance of the Ministry of Education, Culture, Sports, Science and Technology and the Ministry of Economy, Trade and Industry.

1.6.2   Interfaces Clarified by Colloid Science

Interfaces were studied in the field of colloid science in the late 1960s. As a result, an interfacial model was proposed. It is well known in this field that colloidal particles, i.e., aggregates of micelles in a surface-active agent, are of nanometer size. This science treats the dispersion systems where a dispersion phase ranges from 1 to 1000 nm in size. The study of interfacial phenomena in colloid science has had a big influence on theoretical clarification of polymer nanocomposites.

1.6.3   Technology Transferred from Composite Materials

Composite materials have been developed by combining more than two different materials to achieve higher performance than the materials' individual performance. The concept of "composite materials" was inherited from ancient times. There are many composite material applications. For example, mud walls made of soil and straw, and composite bows made of wood and bamboo reinforced with animal bone, sinew and glue are all composites. Modern composite technology has been developed dramatically since Bakelite® (polyoxybenzylmethylenglycolanhydride), one of the phenolic resins developed by L. H. Baekeland in 1907. This can be compounded with reinforcing substances such as wood powder, paper, cotton, and textile. After that, glass fiber–reinforced plastic (FRP) was invented in 1935 and expanded from the special-purpose military market into the more general consumer market. Furthermore, epoxy as the glue was industrialized in 1946, and later epoxy composite filled with inorganic filler was used for structural members. Dispersion filler phases play a role of reinforcing polymers and/or compensate for characteristics insufficient when only polymers are used. As the technology of composite materials became highly developed, the size of dispersion phases became gradually smaller. Composite materials filled with silica or alumina fillers of micrometer size, and polymer alloys developed in the 1990s are now widely utilized in many sectors. As indicated in Fig. 1.2, from the standpoint of the size of filler phases, recently emerging nanocomposite materials have a certain technological continuity with traditional microcomposite materials. In other words, nanocomposites that have succeeded microcomposites in concept have been developed by making dispersion phases finer and finer.

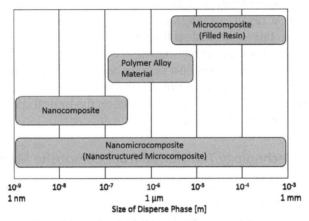

Figure 1.2   Size of disperse phase in composite materials.

1.6.4 Birth of Polymer Nanocomposites

The polyamide/clay nanocomposite was developed in 1987 by the Toyota R&D group. That was the beginning of polymer nanocomposites as engineering plastics. The Toyota R&D group succeeded in fabricating nanocomposite in a way like organically modified clays are melt-compounded, where monomers (ε-caprolactam) were inserted into the interstices between neighboring layers and polymerized by the open-ring polymerizing process. This clay/polyamide nanocomposite has higher mechanical properties and heat resistance than its base resin. It has been in practical use since the 1990s.

1.7 Bright Future of Nanocomposite World

Polymer nanocomposites have been developed to achieve R&D targets such as the establishment of fabrication methods, optimization of filler conditions in polymers, characterization of fabricated nanocomposite dielectrics, comparative computer simulation study, and R&D exploration for practical use. As a result, R&D has made great progress but is still halfway to its completion. In order to promote further progress and move on to the next growing stage, further R&D should be carried out to optimize fabrication methods of nanocomposite dielectrics, clarify polymer/nanofiller interfacial structures and dielectric and insulation phenomena, to deepen computer simulation and explore and expand applications. In the near future, nanocomposites will have not only excellent dielectric properties but also other superb characteristics such as low thermal expansion coefficient, high thermal conductivity, high heat resistance, high mechanical strength, high flame retardancy, high permeability, and high and low permittivity. Thus, nanocomposites are expected to become multi-functional super-composites that possess all of these properties.

An increase in oil demand, shortage of metal resources, and global greenhouse effect are anticipated due to increase in energy use associated with the world population explosion. In order to solve this issue, it is necessary to construct a sustainable recycle-based society. This would be rather difficult without wide-ranging development of science and technology. Technological innovation is also required in the field of material science and technology. As shown in Fig. 1.3, if high-performance nanocomposites become multi-functional super-composites with functions such as recycle, carbon neutral, low environmental load, metal substitute, and biomimetics, they will help to solve these controversial problems.

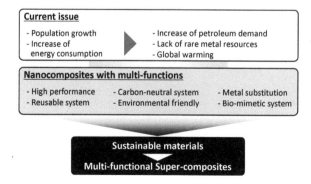

Figure 1.3 Nanocomposite dielectrics for sustainable society.

Further development of nanocomposites requires intensive interdisciplinary studies based on electrical engineering, chemistry, materials science, physics, and computer simulation, and collaborations between the industry and academia. In addition to the famous Richard P. Feynman's words, the following phrase is proposed here: "There's plenty of room for nanocomposite insulating materials!" Please join us for further fruitful development.

Useful Books and Papers for Readers

1. Nelson, J. K. (2010). Burke, J. E. (1964). *Dielectric Polymer Nanocomposites*, 1st ed. (Springer), pp. 1–368.
2. Nakajo, M. (2003), *Polymer Nanocomposites* (Kogyo Chosakai Publishing), pp. 1–299 (in Japanese).
3. Tanaka, T. and Imai, T. (2013) Advances in nanodielectric materials over the past 50 years, *IEEE EI Magazine*, **29**(1), pp. 10–23.

PART I
APPLICATIONS

Chapter 2

Potential Applications in Electric Power and Electronics Sectors

Takahiro Imai[a], Yoshiyuki Inoue[b], Masayoshi Nagata[c], Tetsuo Yoshimitsu[d], Kazutoshi Goto[e], Takanori Kondo[f] and Takashi Ohta[g]

[a]*Toshiba Corporation*
[b]*Sumitomo Electric Industries, LTD.*
[c]*University of Hyogo*
[d]*Toshiba Mitsubishi-Electric Industrial Systems Corporation (TMEIC)*
[e]*Technology Consultant*
[f]*NGK INSULATORS, LTD.*
[g]*Panasonic Corporation*

2.1   Power Apparatus and Cables

Takahiro Imai[a] and Yoshiyuki Inoue[b]

[a]*Power and Industrial Systems R&D Center, Toshiba Corporation,*
Fuchu-shi, Tokyo 183-8511, Japan
[b]*Technology Development Department Electric Wire & Cable, Energy Business Unit,*
Sumitomo Electric Industries, LTD.,
Hitachi-shi, Ibaraki 319-1414, Japan

takahiro2.imai@toshiba.co.jp, inoue-yoshiyuki1@sei.co.jp

Polymer nanocomposites have been developed for practical use in electric power applications. For example, epoxy-based nanomicrocomposites have been developed for solid insulated

---

Advanced Nanodielectrics: Fundamentals and Applications
Edited by Toshikatsu Tanaka and Takahiro Imai
Copyright © 2017 Pan Stanford Publishing Pte. Ltd.
ISBN 978-981-4745-02-4 (Hardcover), 978-1-315-23074-0 (eBook)
www.panstanford.com

switch-gear (SIS). This kind of switchgear is environmentally friendly and uses solid insulation systems instead of sulfur hexafluoride (SF_6) gas insulation systems. Moreover, cross-linked polyethylene nanocomposite has been developed to prevent space charge accumulation and is used in insulation systems in DC power cables.

2.1.1 SF_6 Gas Has Excellent Performance but Also Has Greenhouse Effect

Electric energy is essential for our everyday lives, and is widely used in lighting, motors, heating and electronic information systems. In advanced countries such as Japan, electricity consumption has increased only slightly in the last decade. However, it has increased rapidly in China, where it is currently about three times what it was 10 years ago [1]. Annual per-capita electricity consumption in 2009 was 7,833 kW-h in Japan and 12,884 kW-h in the USA, while the global average was 2,730 kW-h and that in China was 2,631 kW-h. This implies that electricity demand will increase dramatically in developing countries such as China in the near future.

Transformers for voltage conversion and switchgears for interruption play an important role in providing steady electricity supplies. These apparatuses are arranged in distribution grids from power plants to customers and utilize the following insulation systems:

(i) Gas insulation systems (Compressed air, SF_6 gas, etc.)
(ii) Oil insulation systems (Paraffinum liquidum, silicone oil, etc.)
(iii) Solid insulation systems (Porcelain, epoxy resin, etc.)

In particular, SF_6 gas has better insulation properties than air. As a result, compressed SF_6 gas contributes to downsizing of transformers and switchgears. SF_6 gas insulated apparatuses are used in distribution systems of more than 22 kV, as shown in Fig. 2.1 [2].

However, Table 2.1 shows that SF_6 gas has 22,200 times the global warming potential (GEP) of carbon dioxide (CO_2) gas. Therefore, the Kyoto protocol on global warming designated SF_6

gas as a greenhouse gas in 1997. This protocol was issued in 2005 after Russia approved its ratification in 2004.

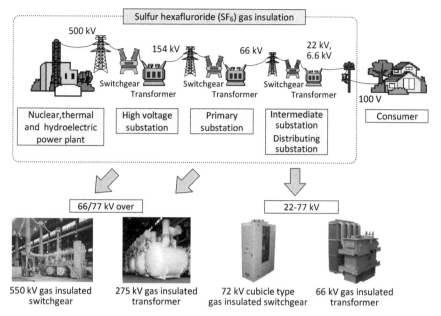

Figure 2.1 Simplified picture of electric power transmission and distribution system [2].

Table 2.1 Global warming potentials of greenhouse gases [3]

Greenhouse gas	Global warming potential (GEP)	Use application/ Emission source
Carbon dioxide (CO_2)	1	-Burning of fossil fuel
Methane (CH_4)	23	-Rice cultivation -Intestinal fermentation of farm animals -Landfill of waste products
Dinitrogen monoxide (N_2O)	296	-Burning of fossil fuel -Industry process

(*Continued*)

Table 2.1 (*Continued*)

Greenhouse gas		Global warming potential (GEP)	Use application/ Emission source
Ozone-unfriendly fluorocarbons	-Chlorofluoro carbon (CFC) -Hydrochloro fluorocarbon (HCFC)	From a few thousand to 10,000	-Spray -Refrigerant of air conditioner and refrigerator -Cleaning process of semiconductor
Ozone-compatible fluorocarbons	-Hydrofluoro carbon (HFC)	From a few hundred to 10,000	-Spray -Refrigerant of air conditioner and refrigerator -Manufacturing process of chemical materials
	-Perfluoro carbon (PFC)	From a few thousand to 10,000	-Cleaning process of semiconductor
	-Sulfur hexafluroride (SF_6)	22,200	-Electrical insulation medium

Note: Global warming potential depends on life of greenhouse gas. -Global warming potential is based on Third Assessment Report 2001, Intergovernmental Panel on Climate Change.

Gas emissions into the atmosphere from SF_6 gas insulated power apparatuses are strictly controlled at present. These apparatuses still play important roles in the transmission and distribution of electric power. However, the electric power industry is positively working to reduce SF_6 gas use. Aging power apparatuses installed in Japan during the high economic growth period will need to be replaced. Moreover, with recent rapid economic growth in overseas countries such as China, electric power demand will increase. Therefore, environmental technology has attracted primary attention in recent developments of new power apparatuses. In the near future, environmental-friendly power apparatuses will play an important role in transmission and distribution of electric power.

2.1.2 Environmentally Friendly Apparatuses Are Developed without Use of Greenhouse Effect Gases

Power apparatuses using compressed dry air insulation systems and solid insulation systems instead of SF_6 gas insulation systems are being developed.

The insulation strength of dry air is approximately one third that of SF_6 gas. Thus, power apparatuses using dry air need to be large due to the longer insulation distances. However, use of compressed dry air enables apparatuses to be downsized. 72/84 kV switchgear using compressed dry air insulation systems have been developed that are the same size as switchgear using SF_6 gas insulation systems, as shown in Fig. 2.2 [4]. The gas pressure of conventional SF_6 gas insulated switchgear is 0.25 MPa (absolute pressure). However, the gas pressure of compressed dry air insulated switchgear is 0.55 MPa (absolute pressure). Moreover, combination of compressed dry air insulation and an insulating barrier using solid material, called "hybrid insulation," decreases the dry gas pressure in 72 kV switchgear, as shown in Fig. 2.3 [5]. The gas pressure decreases to approximately 0.3 MPa (absolute pressure).

 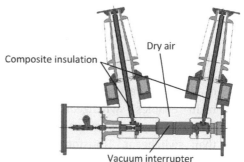

Figure 2.2   72/84 kV compressed dry air insulated switchgear [4].

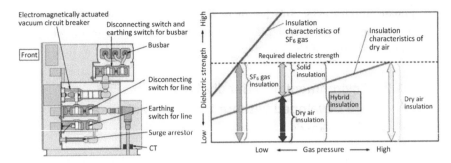

Figure 2.3 72 kV compressed dry air insulated switchgear using insulating barrier technology [5].

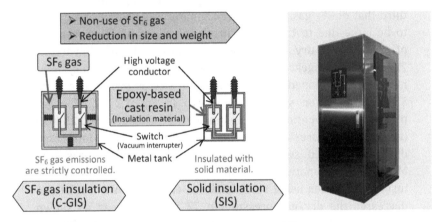

Figure 2.4 24/36 kV solid insulated switchgear [6].

24/36 kV switchgear using a solid insulation system has been developed as shown in Fig. 2.4 [6]. This SIS is composed of a vacuum interrupter and a metal conductor molded with cast epoxy resin containing SiO_2 microfillers. Generally, the insulation strength of cast epoxy resin is approximately three times that of SF_6 gas. Therefore, a solid insulation system using cast epoxy resin enables switchgear to be downsized. Moreover, solid insulation systems are also used in power apparatuses other than switchgear. Transformers insulated with cast epoxy resin instead of oil and SF_6 gas insulation have also been developed.

2.1.3 Nanofiller Dispersed Insulating Materials Are Being Developed for Power Apparatuses

Epoxy resin has superior insulation properties. It was first used in power apparatuses in the 1940s and is still commonly used.

However, if they contain defects such as voids, their insulation properties severely decrease due to degradation such as electrical treeing and partial discharge of epoxy resin. Therefore, epoxy resin with superior insulation properties under high voltage is needed to expand the use of solid insulation systems instead of SF_6 gas insulation systems. Nanofiller dispersion is an effective approach to improving the insulation properties of epoxy resin. Some developments of epoxy resin containing nanofillers towards practical use described as follows.

2.1.3.1 Development of nanomicrocomposite for solid insulated switchgear

In solid insulation switchgear, high-voltage parts such as vacuum interrupters and metal conductors are molded with epoxy resin. Exfoliation occurs at boundaries between high-voltage parts and epoxy resin due to heat in operation. Therefore, high loading of inorganic microfillers decreases the thermal expansion of epoxy resin to the level of metal conductors. In this conventional filled epoxy resin, SiO_2 and Al_2O_3 are generally used as inorganic microfillers, and filler content is more than 60 wt%.

Nanofiller dispersion improves the insulation properties of conventional filled epoxy resin. Figure 2.5a shows an electron micrograph of epoxy resin containing 64 wt% SiO_2 microfillers and a few wt% clay nanofillers. This is called a nanomicrocomposite. The distance between fillers in a nanomicrocomposite is one fifth that of a conventional filled epoxy that has the same amount of SiO_2 microfillers. The picture demonstrates that nanomicrocomposite has a densely-packed structure of fillers.

Moreover, under constant AC voltage using the needle-plate electrode configuration, this nanomicrocomposite shows 24 times longer insulation breakdown time than conventional filled epoxy resin that has 65 wt% SiO_2 microfillers, as shown in Fig, 2.5b. Figure 2.5c estimates the mechanism of improvement in which the dispersed nanofillers and microfillers seem to effectively prevent electrical tree propagation. In addition, a component model of SIS composed of a connecting conductor, a vacuum interrupter and a base conductor insulated with nanomicrocomposite was made for practical use as shown in Fig. 2.6 [7].

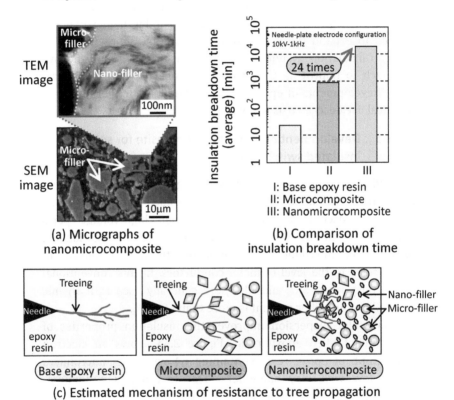

Figure 2.5 Structure and improvement of resistance to electrical tree in nanomicrocomposite [2].

2.1.3.2 Development of combined insulation system for molded instrument transformer

Switchgear has an instrument transformer (IT) for transforming voltage to an adapted low value for measurement. It is composed of an iron-core, coils, and epoxy molding material. In the molded type IT, the combined insulation system, which means a combination of enameled wire with nanocomposite coating and a nanomicrocomposite, is developed to give it longer-life operation without insulation breakdown between the layers in the molded coils [8].

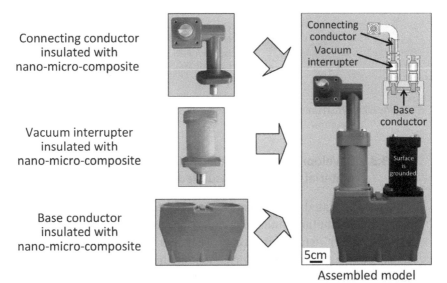

Figure 2.6 Component model of solid insulated switchgear (SIS) insulated with nanomicrocomposite. © 2010 IEEE. Reprinted, with permission, from Imai, T., Yamazaki, K., Komiya, G., Murayama, K., Ozaki, T., Sawa, F., Kurosaka, K., Kitamura, H., and Shimizu, T. (2010). Component Models Insulated with Nanocomposite Material for Environmentally-Conscious Switchgear, *Proc. IEEE ISEI*, pp. 299–232.

A model specimen composed of nanocomposite enameled wire and nanomicrocomposite was prepared to measure the average insulation breakdown time, as shown in Fig. 2.7. The model specimen shows five times longer breakdown time than the specimen with a conventional insulation system without nanofillers [9]. Moreover, the making of this trial model demonstrates the feasibility of an IT molded with nanomicrocomposite.

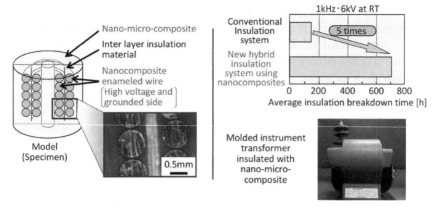

Figure 2.7 Instrument transformer insulated with nanomicro composite [9].

2.1.3.3 Development of high-thermal-conductivity insulation material for all solid insulated substation

SF_6 gas insulated switchgears and oil-filled transformers are used in distribution and interconnection substations. Increasing electric power demand in urban areas and replacement of aged apparatuses need new environmentally-friendly and compact facilities. Therefore, an all solid insulated substation using a solid insulation system is proposed to fulfill this demand, as shown in Fig. 2.8a [10].

In SF_6 gas insulated and oil-filled transformers, the SF_6 gas and oil work as a circulator to release heat generated in the iron-core coil as well as in the insulation medium. Therefore, a transformer using an all solid insulation system needs both high thermal conductivity and insulation properties that are superior to those of conventional insulation materials.

Figure 2.8b shows the both properties of an epoxy-based composite containing aluminum nitride (AlN) spherical nano-micro-hybrid fillers prepared by the transferred type arc plasma method. This filler has a unique structure in which the AlN nanofillers attach to the surface of AlN microfiller. This composite has both high thermal conductivity and sufficient AC insulation breakdown strength for an all solid insulated transformer [11].

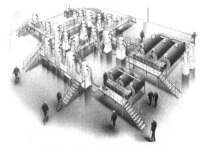

(a) All solid insulated substation

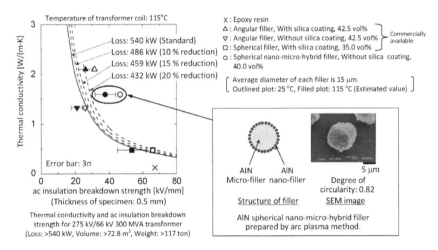

(b) Estimate of Insulation material for all solid insulation substation

Figure 2.8 Proposal of all solid insulated substation [10, 11].

2.1.4 DC Is Better Than AC for Long-Distance Power Transmission

Generally, electric power is transmitted as AC current. However, DC transmission becomes advantageous for long-distance transmission, for example, by submarine cables. One reason for this is that DC has smaller loss during power transmission. During AC transmission, energy loss occurs due to both Joule heat generation in the conductor and dielectric loss of the insulator, while DC transmission loses energy due to the former alone. Dielectric loss in AC transmission is affected by various factors, for instance, electric field, insulation material, temperature, etc. When using cross-linked polyethylene (XLPE) for insulation, which is most often used in underground power transmission lines, power transmission at voltages above 100 kV, energy loss due to dielectric loss in the insulation and that due to Joule heat generation in the conductor becomes approximately the same. However, DC power transmission usually requires AC-DC conversion, thus incurring cost of conversion and conversion equipment. Therefore, economic benefits of DC power transmission are impaired in short distance transmission.

Transmission capacity can be cited as a benefit of the latter. In AC power transmission, in addition to effective power transmission by the current (watt current), quadrature current is generated. Quadrature current, considering the power cable as a capacitor formed by the conductor–insulator–outer shielding layer (ground potential), can be considered as a current portion that is used for charging and discharging the capacitor. If transmission distance is increased, the ratio of quadrature current to watt current is increased. Therefore, the amount of power available to the power receiving side is reduced. The decrease ratio depends on the insulation material, transmission voltage, and transmission distance. For example, using XLPE insulated power cables to transmit 50 km at a voltage of 500 kV, watt current is reduced by approximately 70%. However, with direct current, the distance does not influence the transmission capacity. As described above, in long-distance power transmission, DC transmission has advantages.

Another advantage is that when an accident occurs in one system, it does not spread to other systems, because they are disconnected at the AC-DC converter. Thus, we can avoid a chain reaction of trouble in a power supply system.

Furthermore, we can keep the whole system stable even if there are fluctuations of the power generation amount, such as when solar and wind power are combined with high power generating systems. Especially, DC transmission becomes more effective in offshore wind power generation, which is being actively introduced in Europe as a renewable natural energy source, as their locations have been shifted from coastal areas to offshore areas due to space constraints.

2.1.5 Space Charge Accumulation Is a Big Nuisance for DC Transmission Cables

Conventionally, oil-impregnated insulation cables, such as mass impregnated (MI) cable and oil-filled (OF) cable, have been applied to DC power transmission. In recent years, with increasing environmental awareness, solid insulated cables with no risk of oil leakage have been developed.

Development of solid insulated DC cables in Japan started in the 1970s. Long-term DC voltage tests on AC-XLPE cables showed that they have many problems for DC applications, including poor breakdown strength at high temperatures, especially with polarity reversal operation. This is due mainly to the effect of space charge accumulation in the insulator [12]. Thus, basic study of solid insulation materials started in the 1980s, and materials such as XLPE nanocomposites, described in the next section, were developed.

The world's first HVDC solid insulation cable was applied to an 80 kV DC line in 1999. Here, a voltage source converter (VSC) using an IGBT (insulated gate bipolar transistor) was used as an AC-DC converter in place of a conventional Line-Commutated Current-Sourced Converter (LCC) using a thyristor. As a VSC does not require inversion of voltage polarity (polarity reversal) when reversing the direction of a power flow, the problem of a decrease in insulation performance, such as space charge influence, can be reduced. VSC converters were initially a little inferior in capacity, conversion efficiency, and breakdown strength, compared to LCC converters. However, these properties have been gradually improved and the number of DC solid insulated cable applications with VSC converter are increasing.

2.1.6 Nanocomposite XLPE Insulated Power Cables Have Been Developed

DC-XLPE material with inorganic nanofiller in XLPE (hereafter denoted as DC-XLPE) were developed from a basic study of solid insulation materials started in the 1980s. Some properties of DC-XLPE are shown in Figs. 2.9 and 2.10, and are compared with those of ordinary XLPE used for AC power transmission (hereafter, AC-XLPE) [13–14].

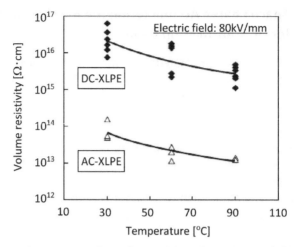

Figure 2.9 Temperature dependence of the volume resistivity [13, 14].

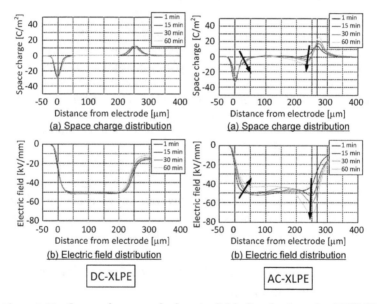

Figure 2.10 Space charge and electric field distributions in DC-XLPE and AC-XLPE at 50 kV/mm, 30°C [13, 14].

Figure 2.9 shows the temperature dependence of volume resistivity measured using sheet samples 150 μm thick. DC-XLPE possesses 100 times higher volume resistivity than AC-XLPE within the range of the measurement temperature.

Figure 2.10 shows space charge and electric field distributions in DC-XLPE and AC-XLPE when a DC voltage of 50 kV/mm was applied using sheet samples 200–300 μm thick. In AC-XLPE, as time progressed, a space charge accumulated inside the specimen, and the electric field distribution became greatly distorted. In DC-XLPE, no space charge was accumulated, and the electric field distribution hardly changed over time.

Figure 2.11 shows the time before the breakdown of sheet specimens about 200 μm thick under a constant DC voltage. It shows that DC-XLPE has a longer life under DC voltage than AC-XLPE. Figure 2.12 shows the breakdown strength of a model cable insulated with DC-XLPE [15] compared with previous data for AC-XLPE cables [16]. The model cable has 9 mm insulator thickness and 200 mm^2 conductor size. The figure shows that the DC breakdown strength of DC-XLPE cable is more than twice that of AC-XLPE cable.

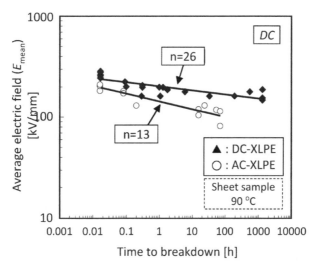

Figure 2.11 DC *V-t* characteristics of DC-XLPE and AC-XLPE at 90°C [13, 14].

The phenomena occurred with the addition of nanoparticles of inorganic filler, and has been researched mainly for low-density polyethylene by many others. With increasing volume resistivity, the restraining of space charge accumulation and similar effects have been reported [17–18].

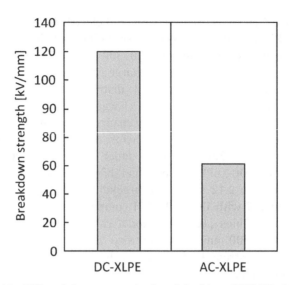

Figure 2.12 DC breakdown strength of model cable at 90°C [13, 14].

The effects of nanoparticles of inorganic filler are not clear, but some hypotheses, for example that nano-sized filler makes a deep trap site and captures the charge [19], have been proposed. The injected charge is captured by the trap near the electrode and remains there. This trapped charge tends to cause relaxation of the electric field near the electrode, which suppresses further injection of charge and limits space charge formation. As shown in Fig. 2.9, with the effect of increasing volume resistivity, the effect of limiting space charge formation should improve DC breakdown strength. The effect of increasing volume resistivity should also prevent thermal destruction by controlling the joule heat.

Practical use of ±250 kV DC-XLPE cable for the HVDC link (45 km length) between Hokkaido-island and Honshu-island in Japan started in December 2012. Figure 2.13 shows the external appearance of the submarine cable. When operation commenced, the operating voltage (±250kV) of this cable was the highest in the world for a transmission line using cables insulated with extruded polymer. And it is the first power transmission cable using nanocomposite material for insulation. It is also the first time a DC extruded cable has been used anywhere for an LCC system including polarity reversal [20].

Figure 2.13 +/- 250kV DC-XLPE submarine cable for the Hokkaido-Honshu DC link line [13, 14].

References

1. The Federation of Electric Power Companies of Japan (2012). *Nuclear and Energy Drawings, Information Library*, p. 1-1-9, p. 1-1-10 (in Japanese).

2. Imai, T. (2006), Nanocomposite Insulation Material for Environmentally-Friendly Power Electric Apparatus, *Toshiba Rev.*, **61**,(12), pp. 60–61 (in Japanese).

3. The Japan Center for Climate Change Actions (http://www.jccca.org), *Properties of Green House Gases*, p. 1–2 (in Japanese).

4. Matsui, Y., Saitoh, H., Nagatake, K., Ichikawa, H., and Sasaki, M. (2005). Development of Eco-Friendly 72/84 kV Vacuum Circuit Breakers, *Proc. IEEJ ISEIM*, No. P2-39, pp. 679–682.

5. Maruyama, A., Takeuchi, T., and Koyama, K. (2009). Dry Air Insulated Switchgear, *Proc. The 40th Symposium on Electrical and Electronic Insulating Materials and Applications in Systems*, No. SS-11, pp. 259–260 (in Japanese).

6. Shimizu, T. (2004). Material Technology for Solid Insulated Switchgear, *EINA (Electrical Insulation News in Asia) Magazine*, pp. 41–42.

7. Imai, T., Yamazaki, K., Komiya, G., Murayama, K., Ozaki, T., Sawa, F., Kurosaka, K., Kitamura, H., and Shimizu, T. (2010). Component

Models Insulated with Nanocomposite Material for Environmentally-Conscious Switchgear, *Proc. IEEE ISEI*, pp. 299–232.

8. Nakamura, Y., Yamazaki, K., Imai, T., Ozaki, T., and Kinoshita, S. (2011). Longer Lifetime of Epoxy/enamel Composite Insulation Systems with Nanocomposite Materials Application, *IEEJ the Paper of Technical Meeting on Dielectrics and Electrical Insulation*, No. DEI-11-088, pp. 65–70 (in Japanese).

9. Imai, T., Cho, H., Nakamura, Y., Yamazaki, K., Komiya, G., Murayama, K., Sawa, F., Ozaki, T., and Shimizu, T. (2012). Nanocomposite Insulating Materials Leading to Product Developments, *Proc. The 43th Symposium on Electrical and Electronic Insulating Materials and Applications in Systems*, No. SS-6, pp. 261–262 (in Japanese).

10. Shibuya, M., Okamoto, T., Kuzuma, Y., Gouda, Y., Suzuki, H., Iwata, M., Kado, H., Hori, Y., Thuchida, H., Kanegami, M., Takahashi, T., Takeda, T., Tanaka, S., Hurukawa, S., and Mizutani, Y. (2000). Proposition of All Solid Insulated Substation, *Central Research Institute of Electric Power Industry (CRIEPI), Research Report*, W00047 (in Japanese).

11. Iwata, M., Furukawa, S., Mizutani, Y., Adachi, K., and Amakawa, T. (2006). Design and Evaluation of All Solid Transformer (Part 4)-Thermal Conductivity and Breakdown Strength of Epoxy Resin with Spherical Nano-structured Composite Particles of Aluminum Nitride -, *Central Research Institute of Electric Power Industry (CRIEPI), Research Report*, H05008 (in Japanese).

12. Investigation R&D Committee on Transition of DC Cable Technology (1999). Transition of DC Cable Technology and Future Tasks, *Tech. Rep. IEEJ*, No. 745 (in Japanese).

13. Murata, Y., Sakamaki, M., Abe, K., Inoue, Y., Mashio, S., Kashiyama, S., Nishikawa, S., Suizu, M., Watanabe, M., Asai, S., and Katakai, S. (2013). Development of High Voltage DC-XLPE Cable System, *Hitachi Densen*, **32**, pp. 5–14 (in Japanese).

14. Murata, Y., Sakamaki, M., Abe, K., Inoue, Y., Mashio, S., Kashiyama, S., Matsunaga, O., Igi, T., Watanabe, M., Asai, S., and Katakai, S. (2013). Development of High Voltage DC-XLPE Cable System, *SEI Tech. Rev.*, **76**, pp. 55–62.

15. Maekawa, Y., Watanabe, C., Asano, M., Murata, Y., Katakai, S., and Shimada, M. (2001). Development of 500 kV XLPE insulated DC Cable, *IEEJ Trans. Power Energy*, **121**(3), pp. 390–398 (in Japanese).

16. Maekawa, Y., Yamaguchi, A., Sekii, Y., Hara, M., and Marumo, M. (1994). Development of DC XLPE Cable for Extra-High Voltage Use, *IEEJ Trans. Power Energy*, **114**(6), pp. 633–641.

17. TF D1.16.03 CIGRE. (2006). Emerging Nanocomposite Dielectric, *ELECTRA* **226**, pp. 24–32.

18. Nagao, M., Murakami, Y., Murata, Y., Tanaka, Y., Ohki, Y., and Tanaka, T. (2008). Material Challenge of MgO/LDPE Nanocomposite for High Field Electrical Insulation, *CIGRE 2008*, D1-301.

19. Ishimoto, K., Tanaka, T., Ohki, Y., Sekiguchi, Y., and Murata, Y. (2009). Thermally Stimulated Current in Low-density Polyethylene/MgO Nanocomposite–On the Mechanism of its Superior Dielectric Properties-, *IEEJ Trans. Fundamentals Mater.*, **129**(2), pp. 97–102 (in Japanese).

20. Watanabe, C., Itou, Y., Sasaki, H., Katakai, S., Watanabe, M., and Murata, Y. (2014). Practical Application of ±250 kV DC-XLPE Cable for Hokkaido-Honshu HVDC Link, *CIGRE 2014*, B1-110-2014.

2.2 Magnet Wires and Motor Windings

Masayoshi Nagata[a] and Tetsuo Yoshimitsu[b]

[a]*Department of Electrical Engineering, University of Hyogo,*
Himeji-shi, Hyogo 671-2201, Japan
[b]*Rotating Machinery System Division,*
Toshiba Mitsubishi-Electric Industrial Systems Corporation (TMEIC),
Yokohama-shi, Kanagawa 230-0045, Japan

nagata@eng.u-hyogo.ac.jp, YOSHIMITSU.tetsuo@tmeic.co.jp

With the development of power electronics technology toward recent promotion of energy saving, many inverter-fed motors are used in electric vehicles or hybrid vehicles. There is fear that the insulation system of motor windings is broken by partial discharge caused by an inverter surge. As a measure against this, nanocomposite enameled wires with superior partial discharge resistance are expected to come into practical use as next-generation magnet wires. Reasons for their long life will be clarified in comparison with general-purpose enameled wires and the future application to production-model motors for improved quality will be considered [1].

2.2.1 Breakdown of Motor Insulation Due to Partial Discharge by Inverter Surge

In electric vehicles or hybrid vehicles, small-sized and high-efficient inverter systems are used to perform the output control of motors. If a sharp and excess surge voltage is applied between windings of a motor coil due to high-speed switching of a power device, partial discharge is caused and the insulation system is broken. This section describes partial discharge phenomenon that occurs across the gap when an impulse voltage with a surge waveform is applied to a twisted-pair sample imitating coil windings, together with the measurement method for the partial discharge.

If an enameled wire used for motor windings is exposed to partial discharge, the coating of the enameled wire is eroded, eventually leading to insulation breakdown. If an impulse voltage imitating a surge voltage is repeatedly applied to a twisted-pair sample while the impulse voltage is increased, partial discharge starts occurring at a certain voltage. This voltage is defined as a partial discharge inception voltage (PDIV). Figure 2.14 shows a picture (captured by a camera) of emission occurring at a contact part of twisted enameled wires [2]. It can be seen that the emission intensity of partial discharge increases in proportion to an applied voltage (V_a in the drawing) and light emitting section extends to the back side of the enameled wires so as to creep on the surface. Insulation degradation progresses in a wide range not within a small wedge-shaped gap near the contact portion of the twisted pair. Figure 2.15 shows an appearance of general-purpose enameled wires that underwent insulation breakdown in a voltage-applied life test using an inverter surge and AC voltage waveform [3]. In the AC insulation breakdown test shown in the right picture, only one breakdown mark is observed. However, in the inverter surge voltage-applied test shown in the left picture, erosion marks caused by streaky partial discharge can also be seen. In an actual inverter-fed low-voltage random wound motor, when a voltage stress more than a certain level is applied between turns, between a phase and the ground, or between different phases, the corresponding positions are damaged, leading to such insulation degradation of windings. Insulation design and evaluation need to be performed so as to prevent partial discharge from occurring.

Occurrence of partial discharge can be measured by various methods such as detection of high frequency voltage fluctuations or pulse current caused by a discharge phenomenon, light or electromagnetic waves emitted by discharge, or chemical changes of gas caused by a discharge arc. In a conventional AC partial discharge measuring instrument, a measuring technique that supports an impulse voltage having an unmeasurable short rise time is studied and, as a sensor for the measuring technique, a high frequency current transformer (CT), photomultiplier tube, horn antenna, macro-strip (patch) antenna, directional coupler, or the like is used [1].

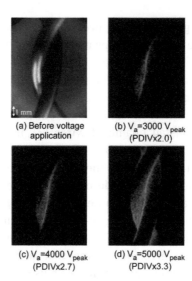

Figure 2.14 Images of emission from the partial discharge occurring at a contact part of twisted enameled wires. © 2006 IEEE. Reprinted with permission from Hayakawa, N., Inano, H., Inuzuka, K., Morikawa, M., and Okubo, H. (2006). Partial discharge propagation and degradation characteristics of magnet wire for inverter-fed motor under surge voltage application, *Annual Rept. IEEE CEIDP*, pp. 565–568.

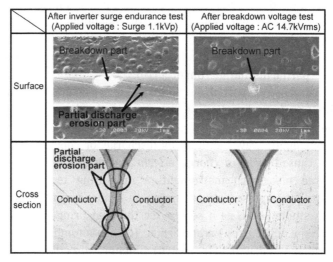

Figure 2.15 Erosion marks appeared on enameled wires that underwent insulation breakdown in a life test using an inverter surge (left) and AC voltage (right) waveform [3].

2.2.2 Partial Discharge Inception Voltage by Inverter Surge Determined by Various Combined Factors

The inception voltage, frequency of occurrence, and intensity of partial discharge are determined by combination of various factors such as the initial electron presence probability, space charge effect, surrounding environment factors (temperature and humidity), various applied voltage waveforms (such as a rise time, frequency, pulse width, and polarity), enameled insulating film characteristics (such as the dielectric constant, type and presence/absence of a filler, and hygroscopicity). It is difficult to accurately predict the partial discharge inception voltage according to these factors. However, when the waveform of an applied voltage is a sine wave, the inception voltage can be estimated based on the voltage applied to the air gap and the curve (Paschen curve) of the Paschen's law. Figure 2.16 shows the evaluated electric field strength together with the Paschen curve in the air by plotting the air gap length $d(x)$ on the horizontal axis [4]. When the voltage V_a is increased, the applied voltage V_a at which the Paschen curve matches the inter-gap voltage is evaluated to be PDIV. In the drawing, when V_a = 915 V, discharge starts when $d(x)$ = 0.03 mm. When V_a = 1500 V, the discharge range extends to the air gap

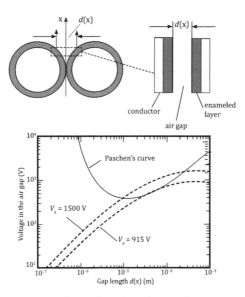

Figure 2.16 Estimation of PDIV based on the Paschen curve in the air [4].

length $d(x) = 0.2$ mm (when the distance x from the enamel film to the contact point changes, the air gap $d(x)$ between enamel films also changes). In the case of a sine wave having a long rise time, the PDIV obtained by experiment well matches this predict value, but the PDIV does not match the predict value in the case of an impulse voltage waveform.

When a twisted pair is simplified as parallel flat plates (see Fig. 2.16), the voltage V_{gap} applied to the air gap is obtained by the following expression [4]:

$$V_{gap} = \frac{V}{\dfrac{2w}{\varepsilon_r d(x)} + 1} \tag{2.1}$$

where ε_r, V, w, and $d(x)$ represent the specific inductive capacity of the enamel insulating film, the applied voltage, the thickness of the enamel insulating film, and the air gap length, respectively. It can be seen from this expression that since changes in the film thickness and specific inductive capacity of an enameled wire have an effect on the effective electric field strength applied across the air gap, the PDIV also changes. Figure 2.17 shows the dependency of the PDIV on the film thickness when a single and positive surge voltage (referred to below as surge A) having a rise time of 120 ns, when a bipolar surge voltage (referred to below as surge B) having a rise time of 300 ns and a repetition frequency of 5000 pps is applied, and when an AC voltage is applied, for a nanocomposite wire having a film in which a nanofiller is dispersed and a general-purpose wire (conventional enameled wire) [5]. As this drawing shows, the PDIV of the surge voltage is higher than that of the AC voltage and increases linearly relative to the film thickness. In this drawing, effects of presence or absence of a nanofiller are not recognized.

Figure 2.18 shows comparison between the specific inductive capacities of two nanofillers (silica and titania) when the filling factor is 10 wt%, 20 wt%, and 30 wt% [6]. In addition, the PDIV characteristics with changes in the filling factor are inspected using a sine wave. Since the specific inductive capacity increases with the filling factor, the PDIV reduces as expression (2.1) shows. Particularly, the specific inductive capacity of titania is slightly larger than that of silica, so the PDIV is relatively low.

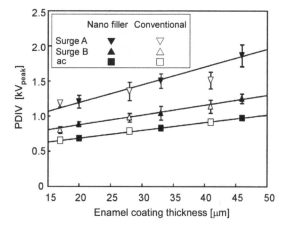

Figure 2.17 Dependency of PDIV on the film thickness of the enameled wires [5].

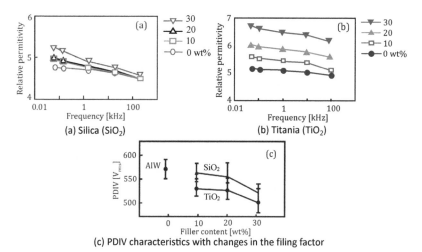

Figure 2.18 Characteristics of the specific inductive capacities of two nanofillers (silica and titania) as a function of frequency and the PDIV with changes in the filling factor [6].

Environmental factors such as temperature or humidity significantly have an effect on the inter-gap electric field and the formation of space charge (electric charge stored after discharge) and significantly change partial discharge characteristics. Therefore, it is thought that temperature and humidity also have an effect on the insulation life of a production-model motor significantly.

Conventionally, partial discharge characteristics have been inspected in an environment in which temperature and humidity are controlled and many experimental results have been reported [1].

Figure 2.19 shows the dependence of the PDIVs of a nanocomposite wire and a general-purpose wire (conventional enameled wire) on humidity [4]. The horizontal axis represents the absolute humidity. The temperature ranges from 30° Celsius to 80° Celsius and the humidity changes from 30% to 95% for each temperature. Although the PDIV is lowered at low temperature (30° Celsius) due to a rise in the relative humidity, the PDIV is hardly lowered at high temperature (80° Celsius). On the other hand, when the relative humidity is 95%, a sharp drop of the PDIV is observed in the case of a general-purpose wire and effects of surface wettability are indicated. Electric charge generated by discharge is stored on the insulating film and, based on the interaction between the electric field formed by the electric charge and the external electric field, the electric field strength applied to the air gap is determined. It is thought that an increase in humidity causes the enamel insulating film to start wetting, the surface conductivity increases, the diffusion of the stored electric charge is promoted, and the PDIV is changed. As a result of surveillance of the temperature and humidity characteristics of the dielectric constant of the enamel insulating film, when the humidity increases, the insulating film absorbs moisture and the dielectric constant of the insulating film increases. Since an increase in the dielectric constant effectively strengthens the electric field strength across the air gap, it is thought that the PDIV is reduced. In addition, since a nanocomposite wire has a different dielectric constant and surface wettability, a nanocomposite wire behaves differently. A modeling study for comprehensively understanding changes in the physical properties of insulating material filled with a nanofiller and changes in a partial discharge phenomenon will become important in the future. Since environmental effects have an influence on the insulation degradation of a production-model motor via changes in a discharge phenomenon, the modeling study is important from a practical viewpoint.

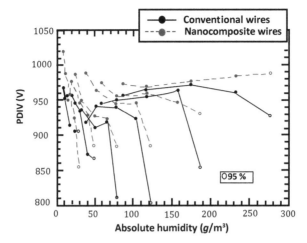

Figure 2.19 Dependence of the PDIVs of a nanocomposite wire and a conventional wire on humidity [4].

2.2.3 Electric Characteristics of Nanocomposite Wire that Significantly Depend on Dispersion State of Nanofiller

A nanocomposite wire has an organic enamel insulating film (such as polyamide-imide, polyester-imide, or polyimide) to which nano-sized inorganic particles (such as silica or titania nanofiller) have been added. To combine excellent partial discharge resistance with a mechanical property capable of standing severe stress in a winding process in a magnet wire, it is important to achieve a uniform dispersion state in which inorganic nanoparticles do not flocculate and an interface state in which a sufficient bond with organic material is ensured. The voltage-applied life characteristics also depend on the manufacturing method, the type and fill amount of inorganic nanoparticles, the insulating film thickness, the use temperature, and the like.

Figure 2.20 shows an example of development concept of a nanocomposite-enameled wire (surge-resistant wire) [7]. Even though the diameter of inorganic particles is small, if the particles flocculate, then charged particles generated by partial discharge selectively enter an organic material portion, cut molecular chains, and advances erosion. If the nanocomposite-enameled wire is bent, cracks may occur. If the filling factor is increased, the ductility of the film is lost, and mechanical properties such as flexibility and adhesiveness are degraded. Therefore, it is important to develop a nanocomposite technique for enabling a uniform dispersion. For example, there is a nanocomposite wire creation technique that can keep a good dispersion state in which inorganic nanoparticles repel each other without being stirred strongly by developing a new nanocomposite manufacturing method that mixes a base resin solution with a colloidal solution in which amorphous silica (with a particle diameter of 20 to 30 nm) is monodispersed in an organic solvent in advance. Two types (polyester-imide base and polyamide base) of surge-resistant enameled wires manufactured by this method are commercially available as KMKED series.

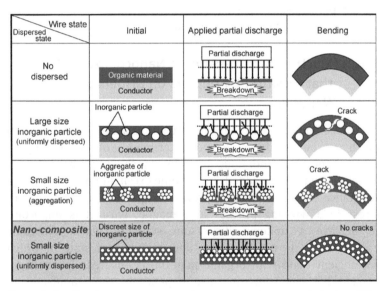

Figure 2.20 Development concept of nanocomposite-enameled wires [7].

In another example, layered silicate (clay) is dispersed uniformly as a nano-sized flat filler using a special shear mixing method [8]. It is said that the orientation of an intercalated flat filler provides effective surge resistance using a small fill amount such as 5 to 10 phr (phr: parts per hundred resin), as shown in Fig. 2.21 [9]. In this method, when the dispersion state of layered silicate is improved, the number of cracks generated in a flexibility test is reduced and voltage-applied life characteristics are improved.

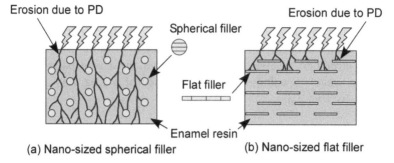

(a) Nano-sized spherical filler (b) Nano-sized flat filler

Figure 2.21 The orientation of an intercalated flat filler to provide effective surge resistance [9].

Surge-resistant wires [10, 11] have been developed in other methods. In one of these methods, a nanocomposite wire is manufactured by solating silica and titania with a solvent and dispersing them in polyamide-imide varnish [10]. There is no large difference in life between silica and titania when filler fill amount is 10 wt%. However, titania has a longer life when filler fill amount is 30 wt%. The life of this surge-resistant wire is approximately 1000 times as long as that of a general-purpose wire when the applied voltage is 2.1 kV$_{peak}$.

2.2.4 Dramatic Improvement of Inverter Surge-Resistant Life by Nanocomposite

It is clear that in application tests of an impulse voltage imitating a surge, the life of a nanocomposite-enameled wire is significantly longer than that of a general-purpose (conventional) enameled wire. The voltage-applied life characteristics (V–t characteristics) indicating the relationship between the time to insulation breakdown and the applied voltage depend on the film thickness of an enameled wire, the type of a nanofiller, and the filling factor.

Figure 2.22 indicates comparison of V–t characteristics between a twisted pair of nanocomposite wires and a twisted pair of general-purpose (conventional) wires [12]. The experiment was performed using a 10 kHz sine wave. As the voltage was reduced, the life to insulation breakdown was significantly extended. Finally, the life became 1000 or more times as large as that of general-purpose wires. Figure 2.23 indicates comparison of $V-t$ characteristics with nanocomposite wires manufactured by filling organic material of polyester-imide with layered silicate and amorphous silica [13]. To evaluate electric characteristics by taking mechanical stresses into consideration, both wires were stretched by 10% and manufactured as twisted pairs. Then, a sine wave AC voltage with a frequency of 1 kHz was applied to the twisted pairs.

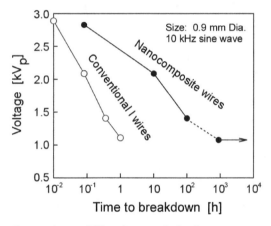

Figure 2.22 Comparison of $V-t$ characteristics between nanocomposite and conventional wires [12].

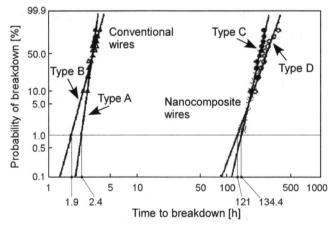

Figure 2.23 Weibull distribution plots of the breakdown time of nanocomposite and conventional wires in the $V-t$ test [13].

As a result, the life of the nanocomposite wires (type C: layered silicate, type D: amorphous silica) was 60 or more times as large as that of conventional wires (type A: polyester-imide wire, type B: polyimide wire).

2.2.5 Mechanism by Which Nanofiller Suppresses Partial Discharge Degradation of Enameled Wire

After voltage-applied life test of the enameled wire, erosion marks caused by partial discharge were seen on the film surface. It is thought that this partial discharge degradation was caused by combination of factors such as collision of charged particles, rises in local temperature, oxidation degradation caused by ozone, effects of space charge. The nanocomposite wire obtains long life characteristics by suppressing such film wastage factors. As shown in Fig. 2.24, if inorganic particles are small and uniformly dispersed, collision of charged particles are prevented by inorganic particles and erosion progresses so as to come behind inorganic particles, thereby increasing the erosion distance [14]. Charged particles undergoing partial discharge are accelerated by the air gap voltage, cut molecular chains of organic material by colliding with the film, and advance erosion. It is thought that erosion is suppressed because the charged particle are reflected and scattered by collision with inorganic nanoparticles (nanofiller) and loses their collision energy.

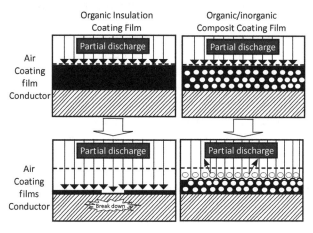

Figure 2.24 Mechanism of erosion suppression by nanofillers [14].

Figure 2.25 is a picture of the wire surface taken after execution of the life test of the nanocomposite wires in Fig. 2.23 with which layered silicate and amorphous silica have been filled [9]. As the voltage application time elapses, surface film by filling nanofiller is gradually formed evenly in the case of layered silicate or formed in a patchy fashion in the case of amorphous silica. It has been found that the precipitated film protects the inside, which is gradually eroded, and contributes to excellent surge resistance.

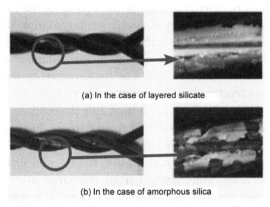

(a) In the case of layered silicate

(b) In the case of amorphous silica

Figure 2.25 A picture of the nanocomposite wire surfaces taken after execution of the life test [9].

2.2.6 Application of Nanocomposite Surge-Resistant Wire to Production-Model Motors and Movement to International Standardization

The excellent characteristics of nanocomposite surge-resistant wires have been evaluated and application to inverter-fed low-voltage random wound motors has already started. However, in the insulation design of conventional low-pressure random wound motors, it is assumed that partial discharge does not occur basically in random wound motors, so nanocomposite surge-resistant wires are used as insurance to prevent insulation breakdown. If nanocomposite surge-resistant wires are introduced in low-pressure random wound motors in the future, since insulation coordination with the inverter drive technology is achieved, the service life can be completed even though partial discharge occurs at a surge voltage, thereby improving the reliability. International movement is currently in progress toward addition as new standards of IEC (International Electrotechnical Commission) [1].

More specifically, IEC/TS60034-18-41 targeted mainly for random wound motors operated at 700 V or lower and IEC/TS60034-18-42 targeted mainly for random wound motors operated at more than 700 V are being discussed as IEC draft standards for inverter-fed motor insulation. IEC/TS60034-18-41 defines that, when a simulated impulse voltage considering an operation voltage including an inverter surge is repeatedly applied after an accelerated test is performed under conditions considering the use environment, repeated partial discharge (five or more times of partial discharge among ten times of application of the simulated impulse voltage) should not occur. IEC/TS60034-18-42 defines that, after a voltage-applied life test is performed, the expected life should be completed. Therefore, IEC/TS60034-18-41 cannot make full use of the advantages of a nanocomposite surge-resistant wire. If IEC/TS60034-18-41 is incorporated in IEC/TS60034-18-42 in the future (that is, an international standard is established so as to allow occurrence of partial discharge in random wound motors operated at 700 V or lower), random wound motors may be manufactured in the procedure as illustrated in Fig. 2.26, thereby enabling the wide spread use of nanocomposite techniques [15].

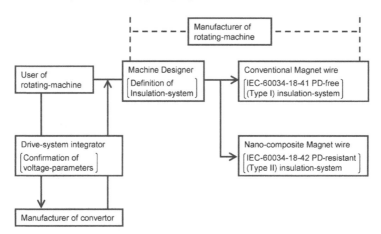

Figure 2.26 Manufacturing procedure of random wound motors expected in the case of establishment of the new IEC [15].

Inverter-fed motors are frequently applied to elevator motors and vehicle motors that concern human life, in addition to those for general industry. In addition, inverter-fed motors are thought to be widely applied to hybrid vehicles, EV motors, or wind power generation in the future on the background of prevention of global warming, so a study to improve the reliability is important.

References

1. Investigating R&D committee on partial discharge measurement under repetitive voltage impulses (2011). Partial Discharge Measurement and Inverter Surge Insulation under Repetitive Voltage Impulses, *IEEJ Technical Report,* vol. 1218.

2. Hayakawa, N., Inano, H., Inuzuka, K., Morikawa, M., and Okubo, H. (2006). Partial discharge propagation and degradation characteristics of magnet wire for inverter-fed motor under surge voltage application, *Annual Rept. IEEE CEIDP*, pp. 565–568.

3. Kikuchi, H., Hanawa, H., and Honda, Y. (2012). Development of polyamide-imide/silica nanocomposite enameled wire, *IEEJ Trans. Fundamentals Mater.*, **132**(3), pp. 263–269 (in Japanese).

4. Kikuchi, Y., Murata, T., Uozumi, Y., Fukumoto, N., Nagata, M., Wakimoto, Y., and Yoshimitsu, T. (2008). Effects of ambient humidity and temperature on partial discharge characteristics of conventional and nanocomposite enameled magnet wires, *IEEE Trans. Dielectrics Electrical Insulation*, **15**(6), pp. 1617–1625.

5. Inuzuka, K., Morikawa, M., Hayakawa, N., Hirose, T., Hamaguchi, M., and Okubo, H. (2006). Partial discharge inception characteristics of nanocomposite enameled wire for inverter-fed motor, *IEEJ The 2006 Annual Meeting Record*, No. 2-054, p. 60 (in Japanese).

6. Hikita, M., Yamaguchi, K., Fujimoto, M., Kozako M., Ohtsuka, S., Ohya, M., Tomizawa, K., and Fushimi, N. (2009). Partial discharge endurance test on several kinds of nanofilled enameled wires under high-frequency AC voltage simulating inverter surge voltage, *Annual Rept. IEEE CEIDP*, No. 7B-20.

7. Kikuchi, H., Yukimori, Y., and Takano, Y. (2005). Properties of inverter surge resistant enameled wire applied organic/inorganic nano-composite insulating material, *The Papers of Technical Meeting of IEEJ*, No. DEI-05-88, pp. 49–54 (in Japanese).

8. Ozaki, T., Imai, T., Sawa, F., Shimizu, T., and Kanemitsu, F. (2005). Partial discharge resistant enameled wire, *Proc. IEEJ ISEIM*, No. A3-9, pp. 184–187.

9. Yoshimitsu, T., Wakimoto, Y., Kobayashi, H., and Ishikwawa, Y. (2007). Consideration on application of pd-resistant-wire for rotating electrical machines, *Proceedings of the 38th Symposium on Electrical and Electron Insulating Materials and Applications in Systems*, No. D-3, pp. 95–99 (in Japanese).

10. Ohya, M., Tomizawa, K., Fushimi, N., Yamaguchi, K., Okada, S., Kozako, M., and Hikita, M. (2009). Voltage-time endurance and partial discharge degradation of nano-composite enameled wire under inverter surge condition, *The 2009 Annual Meeting Record, IEEJ*, No. 2-052, p. 62 (in Japanese).

11. Boehm, F. R., Nagel, K., and Schindler, H. (2007). Voltron TM-A new generation of wire enamel for the production of magnet wires with outstanding corona resistance, *Dupont Catalog*.

12. Kikuchi, H., and Asano, K. (2006). Development of organic/inorganic nano-composite enameled wire, *IEEJ Transactions on Power and Energy*, **126**(4), pp. 460–465 (in Japanese).

13. Uozumi, Y., Kikuchi, Y., Fukumoto, N., Nagata, M., Wakimoto, Y., and Yoshimitsu, T. (2007). Characteristics of partial discharge and time to breakdown of nanocomposite enameled wire, *The Papers of Technical Meeting of IEEJ*, No. DEI-07-64, pp. 51–56 (in Japanese).

14. Kikuchi, H., and Hanawa H. (2012). Inverter surge resistant enameled wire with nanocomposite insulating material, *IEEE Trans. Dielectr. Electr. Insul.*, **19**(1), pp. 99–106.

15. Yoshimitsu, T. (2009). Consideration on nanocomposite magnet wires and surge resistant properties, *The 2009 Annual Meeting Record, IEEJ*, No. 2-S2-8, pp. S2(29)–(32) (in Japanese).

2.3 Polymer Insulators for Outdoor Use

Kazutoshi Goto[a] and Takanori Kondo[b]

[a]Kawasaki-shi, Kanagawa 215-0006, Japan
[b]NGK High Voltage Laboratory, NGK INSULATORS, LTD.,
Komaki-shi, Aichi 485-8566, Japan

gotokazu2@jcom.home.ne.jp, t-kondou@ngk.co.jp

Polymer insulators for outdoor use are lightweight owing to their composite structure. The external layer of silicone rubber is resistant to ultraviolet radiation and erosion. Nanofiller dispersion improves erosion resistance because the interfacial region between nanofiller and silicone rubber enhances thermal resistance to arc discharge. Research on silicone nanocomposites keeps evolving, and this section describes the technology expansion for practical use.

2.3.1 Polymer Insulators Are Lightweight, Composite Structures

Polymer insulators for outdoor use were developed in the United States in the 1960s for distribution of electric power. However, these insulators had insufficient resistance to ultraviolet radiation and insulation, and went out of production. After that, polymer insulators made from epoxy resin, EVA (ethylene-vinyl acetate copolymer) and EPR (ethylene propylene rubber), and EPDM (ethylene-propylene-diene copolymer), with silicone rubber for the external (jacket) material, were developed in the 1980s. At present, silicone rubber is predominately used for polymer insulators for outdoor use [1].

Polymer insulators are composed of silicone rubber for electrical insulation and fiber reinforced plastic (FRP) for mechanical strength, as shown in Fig. 2.27. They are used as suspension, dead-end and inter-phase insulators in transmission lines. In Japan, the use of polymer insulators is limited to experimental applications because porcelain insulators have been mainly used. However, polymer insulators are adopted for inter-phase spacers because of their light weight. An example of an application of an inter-phase spacer for a 275 kV line is shown in Fig. 2.28 [2]. There are also examples of polymer insulators for protection against seismic load.

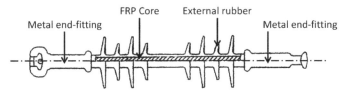

Figure 2.27 Structure of polymer insulator [1].

Figure 2.28 Inter-phase spacer for 275 kV using polymer insulator [2].

Polymer insulators have various advantages such as light weight, easy handling, high resistance to pollution and excellent water-repellent recovery characteristics compared to porcelain insulators. Therefore, polymer insulators and polymer bushings are used in heavy apparatuses to decrease damage due to seismic load. However, polymer materials are inferior to ceramic materials in severe usage environments, but they are easier to manufacture. Thus, it is necessary to take a material's life into consideration when considering the use of polymer insulators.

2.3.2 What Kinds of Performances Are Needed for Polymer Insulators?

Many factors such as dust grains, sea-salt grains, rainfall, dew condensation and exhaust gases from cars and factories (NOx, SO_x, and chemical mist) affect electrical insulation properties of polymer insulators when they are used in outdoor environments. These substances adhere and deposit on the polymer surface, and cause degradation of insulation properties.

In particular, ultraviolet radiation from sunlight has an impact on the degradation of polymer materials. The bond energy of carbon–carbon (C–C) in the main chemical chains of organic materials is smaller than the energy of sunlight, as shown in Table 2.2 [3]. Therefore, most organic materials are degraded by ultraviolet radiation. Silicone rubber, which has the chemical structure shown in Fig. 2.29, is resistant to ultraviolet radiation because the bond energy of silicon–oxygen (Si–O) is larger than the energy of sunlight. Accelerated outdoor weathering tests and light resistance tests demonstrate that silicone rubber is more resistant to ultraviolet radiation than epoxy resin and other rubber materials. Additionally, the hydrophobic property of silicone rubber prevents dust from adhering to the surface of polymer insulators. This contributes to its water-repellent recovery characteristics [4].

Another severe degradation factor is heat damage due to arc discharge generated locally on wet polymer surfaces when polymer insulators are used in outdoor environments [4]. Carbonized conducting paths (tracking) and erosion appear on the polymer surface due to degradation attributed to repetitive arc discharge. These degradation phenomena cause marked degradation of electrical insulation properties of polymer materials. Therefore, polymer insulators need excellent resistance to these phenomena. One effective approach to improving degradation resistance is to use silicone-based polymer nanocomposites, and many researches are being conducted.

Table 2.2 Comparison of bond energy [3]

Bond	Bond energy [kcal/mol]
Si–O	106.0
C-H	98.7
C–C	82.6
Si–C	78.0

*Energy of sunlight (300 nm): 96.6 kcal/mol.

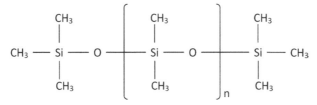

Figure 2.29 Chemical structure of silicone rubber (polydimethylsiloxane).

2.3.3 Erosion Resistance Is Improved by Using Nanocomposites

Silicone rubber, which has a silicon–oxygen bond in the main chemical chain, is resistant to tracking degradation (degradation with formation of carbonized conducting path) compared to other polymer materials.

Degradation due to arc discharge has a strong tendency to erode the polymer surface. Therefore, development of erosion-resistant materials is of primary importance for polymer insulators.

Generally, 50–70 wt% silica fillers and alumina trihydrate (ATH) fillers of micro-meter dimension are infilled in the silicone rubber to improve erosion resistance. However, silicone rubber nanocomposite containing only 2–10 wt% nanofillers shows the same erosion resistance as silicone rubber containing 50–70 wt% microfillers. Moreover, nanofiller dispersion in silicone rubber tends to improve thermal resistance and mechanical strength. Thus, there have been an increasing number of technical papers concerning silicone rubber nanocomposites since around 2004 [5–7].

Figure 2.30 compares the resistance to erosion according to ASTM 2303 (IEC 60587) in room temperature vulcanization (RTV) of silicone rubbers containing 5, 10 fused silica nanofillers or 10, 30, 50 wt% fused silica microfillers [5]. The results demonstrate that silicone rubber containing 10 wt% nanofillers has smaller eroded volume than silicone rubber containing 10 wt% microfillers. In particular, silicone rubber containing 10 wt% nanofillers has smaller eroded volume than silicone rubber containing 50 wt% microfillers, which shows the excellent effect of nanofiller dispersion on erosion resistance. In this test, the polymer surface is wetted by ammonium chloride solution. Joule heat due to leak current vaporizes the solution on the polymer surface, and a dry area (dry-band) appears. The interfacial region formed by the nanofillers seems to impart arc discharge resistance to the dry area (dry-band arc) of silicone rubbers [6].

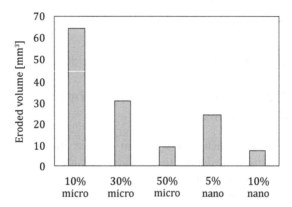

Figure 2.30 Comparison of resistance to arc discharge in silicone rubber containing nanofillers or microfillers [5].

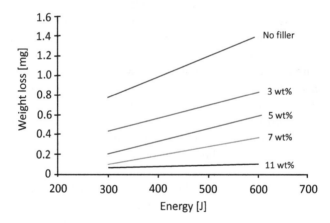

Figure 2.31 Weight loss as a function of laser energy for different nanofilled silicone rubber samples [6].

Weight loss of silicone rubbers due to heat energy of the dry-band arc are estimated in laser irradiation experiments. Figure 2.31 shows weight loss by laser irradiation (802 nm-5W, 60 and 120 second irradiation) in two-part RTV silicone rubber containing 3, 5, 7, 10 wt% hydrophobic fumed silica nanofiller (diameter 12 nm) and silicone rubber without fillers [6]. Erosion resistance is improved in silicone rubber nanocomposites. In particular, silicone rubber nanocomposite containing 10 wt% nanofillers has much smaller weight loss than any other silicone rubbers.

Other approaches such as atom force microscope (AFM) observation, thermo gravimetric analysis (TGA), Fourier transform infrared spectroscopic analysis (FT-IR) and thermal conductivity measurement evaluate the effects of nanofiller dispersion on erosion resistance. These results estimate that the strong chemical bond between nanofillers and silicone rubber at the interfacial region improves resistance to heat and dry-band arc.

2.3.4 Bonding in Interfaces Is Strengthened by Nanofiller Addition

The interface between nanofillers and polymer is much larger than that between microfillers and polymer, as shown in Table 2.3. Some nanofillers have approximately 100 times larger interface areas than microfillers, and approximately four times larger density than microfillers [8]. Nanofiller dispersion in silicone rubber forms large interfacial regions, and seems to prevent silicone rubber from degrading due to heat. Figure 2.32 compares the thermal degradation of silicone rubber containing 2.5 wt% nanofillers or microfillers. Compared to the final residual weight ratio, silicone rubber containing alumina nanofillers, nano natural silica and nano fumed silica have 43%, 55%, and 59% residue, respectively, although silicone rubber containing microfillers has 32% residue.

Table 2.3 Properties of various nanofillers

Filler	Average particle size [nm]	Surface area [m^2/g, BET]	Density [g/cm^3] (at 25°C)
Nano fumed silica (nfs)	7	390 +/− 40	2.2
Nano natural silica	10	590–690	2.2–2.6
Nano alumina	2–4	350–720	4
Micro silica (m)	5000	5	0.58

Source: © 2009 IEEE. Reprinted with permission from Ramirez, I., Jayaram, S. H., Cherney, E. A., Gauthier, M., and Simon, L. (2009). Erosion resistance and mechanical properties of silicone nanocomposite insulation, *IEEE Trans. Dielectr. Electr. Insul.*, **16**(1), pp. 52–59.

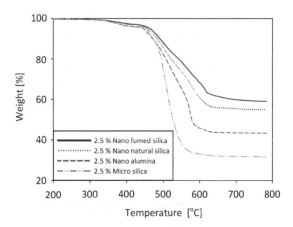

Figure 2.32 Thermal degradation of silicone rubber containing nanofillers. © 2009 IEEE. Reprinted with permission from Ramirez, I., Jayaram, S. H., Cherney, E. A., Gauthier, M., and Simon, L. (2009). Erosion resistance and mechanical properties of silicone nanocomposite insulation, *IEEE Trans. Dielectr. Electr. Insul.*, **16**(1), pp. 52–59.

Each silicone rubber containing nanofillers has much higher residue than those containing microfillers. Moreover, the kinds of nanofillers seem to affect the state of the interfacial region, and the differences between interfacial regions reflect differences of residue. The figure demonstrates that nanofiller dispersion improves resistance to thermal degradation in the silicone rubber nanocomposites. This improvement is attributed to the large nanofiller interface, and the increase in concentration of hydroxyl group on the silica nanofiller surface that participates in the interaction has an impact on the physical interaction in the interfacial region. This interaction is based on the hydrogen bond between the hydroxyl group on the silica nanofiller surface and the siloxane chain of the silicone rubber matrix.

As mentioned above, the interaction between nanofillers and silicone rubber plays an important role. Therefore, it is important to improve nanofiller dispersibility to strengthen the interaction in the interfacial region. Addition of surface-active agent (Triton X-100) improves the dispersibility of nanofillers [8]. Moreover, erosion resistance and mechanical properties in the silicone rubber containing both nano- and microfillers are evaluated. Figure 2.33 shows the eroded mass as a function of

Triton concentration in silicon rubber containing 2.5 wt% silica nanofillers in a laser irradiation test. In this test, the nanofillers are treated with calcination at 300 and 600°C, and the pph (parts per hundred) means the concentration of nanofillers. The eroded mass decreases with increase in Triton concentration. The improvement of nanofiller dispersibility owing to the Triton addition prevents the silicone rubber matrix from degrading due to heat caused by the laser irradiation. However, there is no difference in the eroded mass based on the calcination temperature in the test.

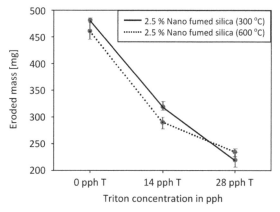

Figure 2.33 Eroded mass due to laser irradiation in silicone rubber containing 2.5% silica nanofillers. © 2009 IEEE. Reprinted with permission from Ramirez, I., Jayaram, S. H., Cherney, E. A., Gauthier, M., and Simon, L. (2009). Erosion resistance and mechanical properties of silicone nanocomposite insulation, *IEEE Trans. Dielectr. Electr. Insul.*, **16**(1), pp. 52–59.

Mechanical properties such as tensile strength, elongation, and hardness are also influenced by the kinds of fillers, amount of fillers and filler/silicone rubber matrix interaction. Table 2.4 shows the mechanical properties of silicon rubber containing nanofillers, microfillers and surface-active agent (Triton). The silicone rubber with Triton addition (2.5% nfs + 14 pph T) has slightly higher tensile strength than silicone rubber without Triton addition (2.5% nfs). The addition of Triton improves nanofiller dispersion in the silicone rubber. However, this result shows that there is no significant effect on tensile strength with Triton addition.

Table 2.4   Mechanical properties of silicon rubber containing fillers

Sample	Ultimate tensile strength [MPa]	Coefficient of variation, CV [%]	Elongation [%]	Hardness (Type A)
2.5% nfs	1.11	15	238	52
2.5% nfs + 14 pph T	1.12	25	235	50
2.5% nfs + 28 pph T	0.99	14	219	50
20% m + 2.5% nfs	2.38	9	214	62
20% m + 2.5% nfs + 14 pph T	2.34	8	209	61
20% m + 2.5% nfs + 28 pph T	2.31	14.9	227	58
20% m + 2.5% nfs + 100 pph T	1.02	33.8	207	47

Source: © 2009 IEEE. Reprinted with permission from Ramirez, I., Jayaram, S. H., Cherney, E. A., Gauthier, M., and Simon, L. (2009). Erosion resistance and mechanical properties of silicone nanocomposite insulation, *IEEE Trans. Dielectr. Electr. Insul.*, **16**(1), pp. 52–59.

The silicone rubber containing both nano- and microfillers (20% m + 2.5% nfs) has more than two times larger tensile strength than the silicone rubber containing only nanofillers (2.5% nfs). However, the effects of Triton addition on tensile strength are not confirmed in the samples of 20% m + 2.5% nfs + 14 pph T and 20% m + 2.5% nfs + 28 pph T.

In particular, tensile strength and hardness dramatically decrease in the sample containing superfluous Triton addition (20% m + 2.5% nfs + 100 pph T). In summary, microfiller dispersion improves the mechanical properties, but optimization of Triton concentration is needed in silicone rubber containing both nano- and microfillers. Furthermore, erosion resistance is also evaluated in these silicone rubber samples [9]. The samples of 20% m + 2.5% nfs + 14 pph T and 20% m + 2.5% nfs + 28 pph T show superior resistance to other samples.

As an approach to understanding the interfacial region, an interphase volume model is proposed [10]. This model estimates the mass formed by the nanofillers and silicone rubber matrix, and shows the correlation between mass and erosion resistance. Experimental results according to IEC 61621 and 60587 show that degradation due to erosion depends on resistance to thermal degradation due to arc discharge. This thermal degradation is not long-term degradation at a certain constant temperature, but localized thermal degradation under high temperature. The strong bond between nanofillers and silicone rubber prevents the silicone rubber matrix from degrading due to localized heat exposure. Moreover, the nanofillers work as a heat barrier. Therefore, silicone rubber nanocomposite shows remarkable erosion resistance.

2.3.5 Technological Evolution Is Expected for Nanocomposite Insulators

Silicone rubber nanocomposites are still in the research stage, and some technical issues remain before they can be put to practical use. Main technical issues are shown as follows:

(i) Homogenous dispersion of nanofillers in silicone rubber.

(ii) Optimization of type and surface state of nanofiller.

Nanofillers make the aggregate agglomerate easily. Thus, issue (i) means that techniques need to be developed to produce silicone rubber nanocomposites containing homogenously dispersed nanofillers. The effects of a different mixer on nanofiller dispersion are reported in the preparation of silicone rubber nanocomposites. The nanofillers are mixed with silicone rubber using a high-speed planetary mixer, a high-pressure homogenizer or a triple-roll mill mixer [11]. Boehmite alumina (AlOOH) is used as the nanofiller, and two-part silicone rubber composed of a resin part and a hardener part are used as the polymer matrix.

The resin part and nanofillers are mixed in a vacuum using the high-speed planetary mixer. The mixture is subsequently additionally mixed with high shear force using the high-pressure homogenizer or triple-roll mill mixer. After that, the mixture is cured in the presence of the hardener. Scanning electron microscopy (SEM) observation of the silicone rubber nanocomposites shows that the triple roll mill mixer is the most effective for dispersing nanofiller homogenously. Moreover, as mentioned above, the addition of a surface-active agent is also effective in improving the nanofiller dispersion state in the silicone rubber nanocomposites.

Issue (ii) means that various considerations concerning the type, surface state and preparation method for nanofiller are needed for silicone rubber nanocomposites containing homogenously dispersed nanofillers. These factors based on the nanofiller have an impact on nanocomposite properties such as erosion resistance, thermal degradation and mechanical strength.

For example, influences of nanofiller preparation methods on thermal degradation properties of silicone rubber nanocomposites are reported [10]. The differences between nanofiller preparation methods are shown in Table 2.5. Figure 2.34 compares the TGA results for the two kinds of silicone rubber nanocomposites containing untreated hydrophilic precipitated silica nanofillers (F1) or untreated hydrophilic fumed silica nanofillers (F2). This figure demonstrates that the nanocomposites containing F1 nanofillers have higher resistance to thermal degradation than those containing F2 nanofillers.

Table 2.5 Differences of nanofiller preparation method

	F1	F2
SiO_2-filler type	**Untreated hydrophilic precipitated**	**Untreated hydrophilic fumed**
Filler particle diameter	20 nm (data sheet)	20–30 nm (TEM images)
Filler content [wt%]	0.5, 1, 2, 5, 10, 15	2, 5, 10

Source: © 2010 IEEE. Reprinted with permission from Raetzke, S., and Kindersberger, J. (2010). Role of interphase on the resistance to high-voltage arcing, on tracking and erosion of silicone/SiO$_2$ nanocomposites, *IEEE Trans. Dielectr. Electr. Insul.*, **17**(2), pp. 607–614.

In the silicone rubber without nanofillers, thermal decomposition begins at 480°C, and the mass decreases rapidly. However, thermal decomposition begins at 560°C in the silicone rubber nanocomposite containing 0.5 wt% F1 nanofillers, and the final residual mass is also large compared to that of the silicone rubber without nanofillers. In particular, the losses of mass due to thermal decomposition are extremely small, and thermal stability is much improved in the silicone rubber nanocomposites containing 5 and 10 wt% F1 nanofillers. Furthermore, the silicone rubber nanocomposite containing 0.5 wt% F1 nanofillers has the same final residual mass as the 10 wt% F2 nanofillers.

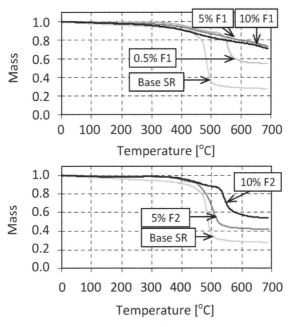

(a) Silicone rubber nanocomposites containing untreated hydrophilic precipitated silica nano-fillers (F1).
(b) Silicone rubber nanocomposites containing untreated hydrophilic fumed silica nano-fillers (F2).

Figure 2.34 Comparison of mass loss in silicone rubber nanocomposites containing nanofillers by different preparation method. © 2010 IEEE. Reprinted with permission from Raetzke, S., and Kindersberger, J. (2010). Role of interphase on the resistance to high-voltage arcing, on tracking and erosion of silicone/SiO_2 nanocomposites, *IEEE Trans. Dielectr. Electr. Insul.*, **17**(2), pp. 607–614.

These results show that the F1 nanofillers are more effective in improving the thermal degradation properties of silicone rubber nanocomposites than the F2 nanofillers, and imply that issue (ii) is very important in nanocomposite preparation. In these silicone rubber nanocomposites, the erosion resistance is also evaluated according to IEC 60587. The silicone rubber nanocomposites containing F1 nanofiller have superior resistance to those containing F2 nanofillers.

The example shows the influences of the preparation methods for silica nanofillers. In the silicone rubber nanocomposites, alumina, boehmite alumina and titania are also used as nanofillers. Moreover, the combined dispersion of nanofiller and microfiller improves erosion resistance [8]. In addition, the optimization of Surface state modified by the surface-active agent is essential for preventing aggregation of nanofillers and dispersing them homogenously.

As stated above, this section describes the polymer insulator and developments of silicone rubber nanocomposites toward practical use in polymer insulators. The excellent erosion resistance of the silicone rubber nanocomposites fascinates many researchers and engineers, although there are some technical issues to be overcome for practical use. Technical innovation regarding silicone rubber nanocomposites is expected in the near future.

References

1. IEEJ Technical Report (2004). No. 948 (in Japanese).
2. NGK Review (1994). No. 54, p. 43 (in Japanese).
3. Barry, J. (1962). *Inorganic Polymers*, Academic Press, p. 244.
4. Investigation Committee on evaluation of discharge property and degradation phenomena of material surface of polymeric insulators (2006). Evaluation of discharge property and degradation phenomena of material surface of polymeric insulators, *IEEJ Technical Report*, No. 1071 (in Japanese).
5. EI-Hag, H., Jayaram, S. H., and Cherney, E. A. (2004). Comparison between silicone rubber containing micro-and nano-size silica fillers, *Annual Rept. IEEE CEIDP*, pp. 385–388.
6. El-Hag, H., Simon, L. C., Jayaram, S. H., and Cherney, E. A. (2006). Erosion resistance of nano-filled silicone rubber, *IEEE Trans. Dielectr. Electr. Insul.*, **13**(1), pp. 122–128.
7. Cai, D., Wen, W., Lan, L., and Yu, J. (2004). Study on RTV silicone rubber/ SiO_2 electrical insulation nanocomposites, *Proc. ICSD 2004*, No. 7P2, pp. 1–4.

8. Ramirez, I., Jayaram, S. H., Cherney, E. A., Gauthier, M., and Simon, L. (2009). Erosion resistance and mechanical properties of silicone nanocomposite insulation, *IEEE Trans. Dielectr. Electr. Insul.*, **16**(1), pp. 52–59.

9. Ramirez, I., Jarayam, S. H., and Cherney, E. A. (2010). Performance of silicone rubber nanocomposites in salt-fog inclined plane, and laser ablation tests, *IEEE Trans. Dielectr. Electr. Insul.*, **17**(1), pp. 206–213.

10. Raetzke, S., and Kindersberger, J. (2010). Role of interphase on the resistance to high-voltage arcing, on tracking and erosion of silicone/SiO_2 nanocomposites, *IEEE Trans. Dielectr. Electr. Insul.*, **17**(2), pp. 607–614.

11. Kozako, M., Higashikoji, M., Hikita, M., and Tanaka, T. (2012). Fundamental investigation of preparation and characteristics of nano-scale boehmite alumina filled silicone rubber for outdoor insulation, *IEEJ Trans. Fundamentals Mater.*, **132**(3), pp. 257–262.

2.4 High-Density Mounted Components for Electronic Devices

Takashi Ohta

Panasonic Corporation,
1048 Kadoma, Kadoma-shi, Osaka 571-8686, Japan

ohta.takashi@jp.panasonic.com

As compactness and high performance of electronic devices progress, there is an increasing demand for high-density mounting of electronic components. Insulation with excellent performance should be explored to satisfy the following requirements. The distance between the conductors is shortened. Die bonding material to bond the encapsulation material is employed to protect the connection of semiconductor elements and the package. Support structures between a semiconductor element and the substrate are devised. BGA (Ball Grid Array) and CSP (Chip-Scale Package) underfill material mounting is sometimes used to keep the interlayer material of the printed circuit board in sound conditions. Moreover, high heat radiation (high thermal conductivity), heat resistance, and electrochemical migration resistance are also a target for intensive research. Polymer nanocomposites are now recognized as novel materials possibly to solve all requirements as stated above.

2.4.1 Polymer Insulators Are of Light-Weight and Composite Structures

In high-performance electronic devices, high-density mounting of components is adopted. Also, in order to cope with an ultra-high frequency band above 10 GHz, these devices employ three-dimensional mounting technologies to avoid the signal delays that may be caused by adding extra wiring routes. In order to meet these needs, electronics devices are designed to shorten the distance between conductors, and between conductors and a build-up substrate. They use, for example, a build-up substrate shown in Fig. 2.35 or a substrate with built-in components shown in Fig. 2.36 [1, 2]. As shown in the figure, the substrate can be made of a ceramic or a resin. However, it might be better to use resin-based active components built in devices such as IC chips. Also, the environments around electronics devices have become more severe. Considering the changes in the requirements, the following measures have been taken. First, high-quality insulation is now under development, since the insulation of multi-layered structures is subjected to high electric stress. The interlayer interfaces are now a focal point for research. Then, compact designs create severe thermal conditions and thus require materials with high heat dissipation properties and high thermal conductivity.

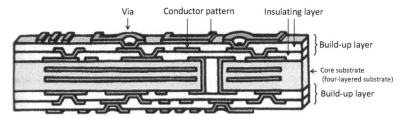

Figure 2.35 Schematic view of build-up substrate [1].

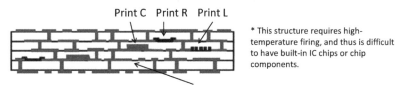

(a) Glass Ceramic Wiring Board with Built-in Passive Components

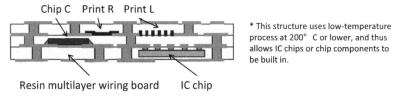

(b) Resin Wiring Board with Built-in Active and Passive Components

Figure 2.36 Schematic view of substrate with built-in components [2].

To cope with the requirements stated above, polymer nanocomposites or polymer nano-microcomposites are considered a new good choice of promising materials for high-density electronics devices. It should be mentioned that materials and insulating substrates to be used for electronics devices are usually subjected to high temperatures during their component mounting processes. For this reason, nano-microcomposites are considered better than nanocomposites. The materials for electronics devices include different kinds of parts, such as parts used at the time of mounting, die-bonding sealing resin, semiconductor elements and a base board, underfill materials, and even capacitor materials as internal parts. The sealing resin is utilized to protect joints in the semiconductor device or in the package from mechanical or thermal stress to the connecting portion of the semiconductor device and the package. The underfill is used when the BGA and CSP are mounted.

For the insulating substrate, high-thermal-conductivity materials are available for substrates or resists as described in Section 2.4.3. The thermal expansion coefficients of these materials should be similar to that of other components in order to prevent mechanical malfunctioning such as warps. Nano-microcomposites fortunately meet these requirements. Furthermore, since the distance between neighboring conductors is extremely short, or for example, from several micrometers to several tens of micrometers, and also the devices have more electrochemical interfaces, electrochemical migration (ECM) and electrical treeing are likely to take place. Nano-microcomposites should have appropriate measures against such detrimental phenomena.

2.4.2 Validity of Nanocomposites as a Sealing Resin of Electronic Components

This section briefly describes nanocomposites and nano-microcomposites for use in the field of information science, or specifically as sealing resins for semiconductor devices and packaging, underfill materials used in semiconductor flip-chip mounting, and capacitor materials.

2.4.2.1 Sealing resins for semiconductor devices and packaging

The heat resistance of semiconductor elements is one of the major R&D items that have been explored in power electronics.

Nanocomposites are also required to have ample heat resistance to ensure sealing resins with high reliability. Conventionally, micrometer-size microfillers (silica fillers) have been used to fill thermosetting resins in order to achieve high heat resistance and low thermal expansion coefficients. However, the microfiller filled thermosetting resins have a drawback: The specific surface of the microfiller used is small, interfacial adhesion between the resin and the microfiller is relatively weak, and they are easily subject to moisture penetration and thereby undergo material degradation. Therefore, since a thermosetting resin with a low thermal expansion coefficient can be possibly fabricated by adding a small amount of the inorganic fillers of nanometer size, such a method is under development.

Sealing resins with low thermal expansion coefficients can be broadly divided into epoxy resins and silicone gels depending on the capacity and the voltage class of the power electronics. The relationship between the device capacity and the voltage is shown in Fig. 2.37 [3]. As shown in the figure, epoxy resins for sealing are used in the voltage range up to 1 kV, while silicone gels are used for higher voltages.

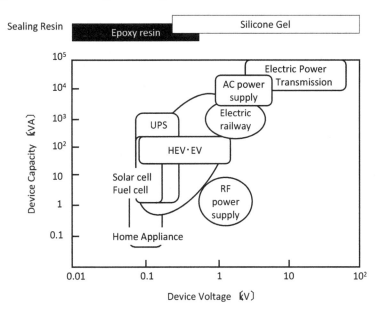

Figure 2.37 Capacity of power electronics and sealing [3].

Figure 2.38 is a schematic cross-sectional view of a power device sealed with a silicone gel [3]. The next-generation devices under development that receive attention as semiconductor elements after silicon are wide-gap semiconductors, such as SiC and GaN (with a band-gap three times as wide as the Si band-gap of 1.2 eV). These materials have high heat resistance and high breakdown strength of about 10 times that of silicon. Wide-gap semiconductors are promising candidates for use in next-generation power devices.

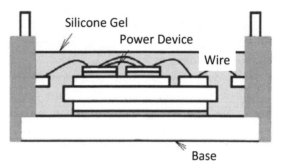

Figure 2.38 Schematic cross section of power device with silicon-sealing structure [3].

Epoxy nanocomposites as sealing materials can be formulated by nano-hybridization in epoxy resins or in silicone gels and have the resultant high heat resistance. Details will be presented below. First, the epoxy resin insulation is composed of nanocomposites complex structures, including whole base materials for electronics devices and various composite parts that have dissimilar interfaces. The insulation is thereby thus not stable to temperature changes and occasionally suffers from mechanical cracks. To avoid this situation, pure epoxy resins are filled with inorganic fillers. Further, to put the high-density mounting to reality, an R&D challenge is to enhance the thermal conductivity of sealing resins.

Thermosetting epoxy resins shown in Fig. 2.39 are filled with organically modified layered silicate clays and then enforced mechanically. The layered silicate clays are homogeneously dispersed and spaced on the nanometer scale between neighboring layers. It has been reported that those composite materials have excellent thermal properties [4].

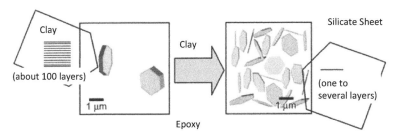

Figure 2.39 Conceptual diagram of nanostructured material clay [4].

The layered silicate clays can be modified by short-length guanidine urea alkyl chain (GU) with multi-functionality. Composite resins filled with the resultant modified clays are expected to exhibit the lowered coefficient of thermal expansion (CTE) and the enhanced glass transition temperature (T_g) compared with pure epoxy resins. Three resins were prepared by adding layered silicate clays modified by short-length guanidine urea (GU) alkyl chain with multi-functionality to epoxy resins and hardening the resins (the selected organic modifiers are shown in Fig. 2.40) [4]. The CTE and the enhanced glass transition temperatures (T_g) of the three composite resins were compared. Table 2.6 shows the results [4].

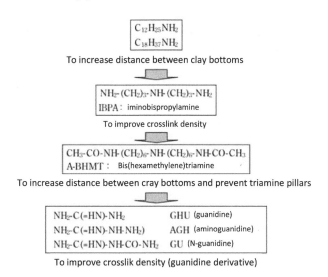

Figure 2.40 Procedure for selecting organic modifiers (guanidine derivative) [4].

Table 2.6 Effects of guanidine urea-modified clays on epoxy resins clay type [4]

Clay type	CTE [10^{-6}/K]	T_g [°C]
Unmodified	63.3	179.7
SiO$_2$ Beads	57.8	181.0
GHU-modified Clay	58.0	176.2
AGH-modified Clay	58.4	177.1
GU-modified Clay	58.1	182.0

Note: Amount of clay dispersion: 5.5 vol%.

As a result, unfortunately, some decrease in T_g is identified in the GU qualified clays. Since the GU includes high heat resistance amino groups, it might be possible to increase or at least maintain the temperature T_g [4]. Therefore, an attempt has been made to lower the coefficient CTE, while maintaining the temperature T_g, by cleaving the clays using a melt-kneading method. This has successfully lowered the coefficient CTE as shown in Table 2.7 [4]. Figure 2.41 is a conceptual diagram illustrating the clay surface. Active hydrogen associated with the amino group of guanylurea acts as a polyfunctional curing agent and contributes to clay cleavage, which results in an increase in the surface area. The short organic chain allows for more damping of molecular vibrations due to interaction between the clay pieces [4]. Another report [5] shows that the size effects influence the flame retardancy of composite materials. In the case of fused silica loaded epoxy resins, the flame retardancy increases with an increase in the filler size.

Table 2.7 Effects of guanidine urea-modified clay mixture on epoxy resins [4]

Clay type	CTE [10^{-6}/K]	T_g [°C]
Unmodified	63.3	179.7
SiO$_2$ Beads	57.8	181.0
Without melt-kneading	58.1	182.0
With melt-kneading	52.6	183.0

Note: Amount of Clay Dispersion: 5.5 vol%.

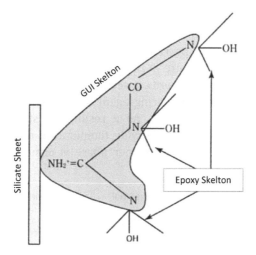

Figure 2.41 Conceptual diagram of clay surface [4].

Next, a silicone gel with polysiloxane, i.e., a special silicone resin with a nano-hybrid structure, is widely used [6]. This resin is highly flexible after curing. It has excellent shock absorption and vibration damping properties and excellent crack resistance. It shows such good performances even in rapid temperature cycling. Integrally molded switching elements with rated voltages of 4.5 KV–120 A are developed. PWM (pulse width modulation) three-phase inverters rated at 110 kV with SIGT (SiC commuted gate turn-off thyristor) are imbedded in the resin [7]. High-temperature composite sealing resins that can withstand temperatures up to 250°C can be fabricated using polysiloxane functional inorganic fine fillers by ultrasonic hybrid technology [8, 9].

2.4.2.2 Die bonding materials for semiconductor flip chips and underfill materials used to fabricate in their mounting semiconductor flip chips

As shown in Fig. 2.42, a film-like adhesive is used to fabricate the semiconductor package, to bond silicon semiconductor chips and substrates, to support structures such as a lead frame, and to conduct in the die bonding processes [10]. The film is used to form acryl/epoxy resin layers. It consists of curing agents, epoxy resins, and acrylic polymers with cross-linkable copolymerized functional groups. Although epoxy resin is uniformly dissolved in the acrylic polymer in the B-stage prior to the curing stage of epoxy resin with a curing agent, the composite might exhibit a clear phase separation structure in the C stage after curing as shown in Fig. 2.43 [10]. In other words, a sea-island structure consisting of the epoxy resin and the curing agent is formed with the acrylic polymer being a sea phase. As a result, the storage modulus decreases significantly over a wide temperature range, while the degradation of the acrylic polymer phase is observed at around 40°C [10].

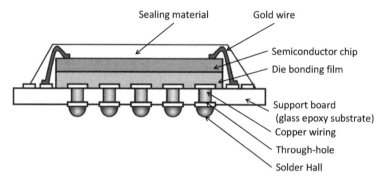

Figure 2.42 An example of a semiconductor package structure [10].

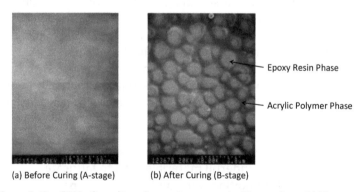

Figure 2.43 SEM of acrylic polymer/epoxy resin film surfaces [10].

The heat resistance of this resin can be improved by incorporating nanofillers. Figure 2.44 shows some of the results [10]. As shown in the figure, the addition of the filler of small size by 4.5 vol% induces a big increase in the tensile elastic modulus at high temperatures. The tensile elastic modulus increases further as the average particle diameter of the filler becomes smaller. Among others, the tensile modulus exhibits a large improvement, for example, an increase by 0.02 μm. The tearing strength of the composites is also larger than that of the unfilled resins as shown in Fig. 2.45. A next problem is associated with underfill materials. These materials are used to mount the flip-chip components for structural reinforcement by filling the gap between the chip and the substrate. Although the resin material is required to be impregnated in a narrow gap in a short time in a narrow gap, no void formation or filler sedimentation is allowed, and excellent adhesion to various substrates is expected.

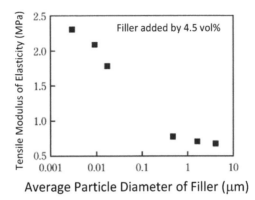

Figure 2.44 Particle diameter of filler vs. tensile modulus of elasticity of film containing filler at high temperature [10].

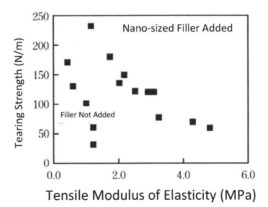

Figure 2.45 Particle diameter of filler vs. tearing strength at high temperature [10].

In order to make high-density mounting of flip-chips feasible, module techniques have been adopted in some packages such as digital cards. Different kinds of materials are used for substrates in this case. Underfill materials used for this purpose usually meet design requirements for the thermal expansion coefficient and elastic modulus to allow low thermal stress. However, it is required to improve the heat cycle resistance of the underfill materials especially for junctions. In addition, the adhesion between different substrates and retardation of moisture ingress are greatly concerned, especially when the materials are subjected to heat cycling. Good underfill materials have been obtained through intensive R&D. For instance, an epoxy/silica nanocomposite material has been developed in which silica fillers with an average particle diameter of 120 nm are uniformly dispersed by 50 wt% [11]. This material has a viscosity as large as 8000 cps, T_g is 143°C, and CTE has been reported to be 37 ppm/°C.

2.4.2.3 High-dielectric-constant capacitor materials

Built-in substrates components in a printed circuit substrate include multiple-elements such as resistors and capacitors mounted on an LSI. They should have the shortest possible wiring to respond properly to super-high frequency and mitigate the delay of the transmitting signals. Capacitors contain built-in fillers with high-dielectric constants dispersed in resin substrates, where poly-disperse fillers are coated on copper foils. In this case, the methods of dispersion and surface treatment need to be improved. One of the examples of commercially used materials is an epoxy resin composite filled with barium titanate fillers, where the filling factor is about 60 wt%, the dielectric constant is 40 to 50, and the dielectric loss tangent is rather high, ranging from 0.02 to 0.03 [12]. New composite materials are also developed for capacitor use, which are filled with barium titanate in alkali developable photosensitive positive polyimide resins, and negative-type polyimide resins. The relative dielectric constant of such composites is around 50 when they are loaded with fillers by 50 wt%. Since the base resin is polyimide, the composite has relatively high temperatures for heat resistance: 700°C for the former (positive type) and 400°C for the latter (negative type) [13]. Some other resins with similar performance have also been also investigated, such as combined materials of silsesquioxane and cycloolefin resins filled with barium titanate nanofillers ($BaTiO_3$) or alumina fillers [14]. The size and the packing density of fillers are also important. In the case of $BaTiO_3$ nanofillers with particle diameter of 29 nm, the dielectric constant of composite films increases proportionately according to the content of the fine fillers, but only up to 30.

Another trial using the electrophoretic deposition (EPD) method indicates that it is possible to fabricate thin films of $BaTiO_3$ nanofillers with dielectric constant 3 by improving the filling rate [15]. This value is two times that of the composite resins prepared by the conventional spin coating method. Formation of ceramic films on resin substrates is another option for fabricating large-dielectric-constant materials. This should take place at the lowest possible temperatures. An aerosol deposition and impinging method can be used for that purpose. With this method, ceramic powders are wound up in a stream of gas and are injected toward the substrate at high speed. In this way, by overcoming the difficult adhesion of ceramics to substrates, it is possible to form films at room temperature. The reported dielectric constant is around 400 [16].

2.4.3 High-Heat-Dissipation, High-Thermal-Conductivity Nano-Microcomposites as Insulating Substrates for Electronic Equipment

Heat dissipation is improved by using nano-microcomposites that act as insulating film substrates, including solder resist materials with high thermal conductivity.

2.4.3.1 High-heat-dissipation, high-thermal-conductivity insulating substrates

High-performance, small-size electronic devices tend to have higher mounting density year by year. Electronic devices, including those used for driving small electronic devices efficiently, automotive electronics devices, lighting equipment such as LEDs, and power devices, require high heat dissipation and high thermal conductivity. For that purpose, it is required not only to select suitable materials but also to optimize the optimum design of system structures. Figure 2.46 shows an example structure [17]. Thermal stress is a major concern. To reduce it, buffer layers, called thermal stress relaxation layers, are usually provided on the internal components of the substrates. Padding with high thermal conductivity, or a conventional large thermal "vias" (filled with thermally conductive material), is used for high-power IC chips to radiate heat outside. The heat transferred to the inner wiring parts from high-power IC chips is radiated to the lower substrate transmitted by the thermal vias. Heat dissipation measures may be modified when thick copper foils are used.

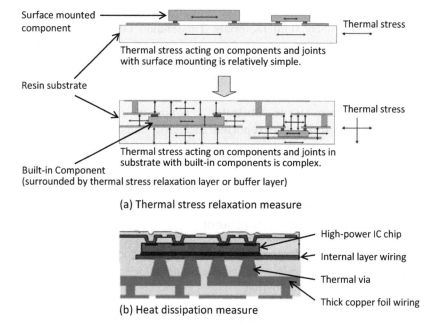

Figure 2.46 Schematic structure with thermal stress relaxation and heat dissipation measures [17].

Nanocomposites or nano-microcomposites are a possible solution for achieving high heat dissipation as they are expected to meet the needs of high-thermal-conductivity insulating substrates, and thus need high-thermal-conductivity inorganic fillers to be dispersed in polymer resins. First, if the resin materials with easily self-assembled structures such as monomesogen type biphenyl groups as shown in Fig. 2.47 are used, it is possible to control higher order structures [18]. The self-alignment characteristics of mesogens are derived from many kinds of microscopic crystal structures and high order anisotropy. The resultant macroscopic structures allow thermally curing reactions from the random state to the fixed stabilizing state. No domain exists in the formed crystal structures. They are independent of each other and are connected by covalent chemical bonds. As a result, phonon scattering reduces at the interface. This is important for obtaining high-thermal-conductivity composites. The thermal conductivity increases to 0.96 W/(m·K), compared with 0.19 W/(m·K) for conventional general-purpose epoxy resins.

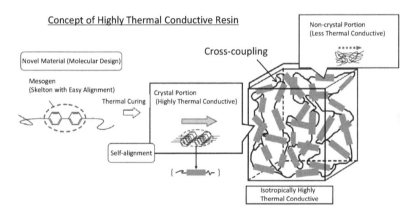

Figure 2.47 High heat conduction of resin with high-order structure control [18].

It has been demonstrated thus far that such low-thermal-conductivity resins can be converted to highly thermally conductive composites by using high-thermal-conductivity inorganic fillers. Further improvement is anticipated. As a specific example, a high thermal conductivity value as high as 10 W/(m·K) is obtained for composites of biphenyl type epoxy with a thermal conductivity of 5 W/(m·K) when loaded by with alumina fillers. This value is close to that of polycyclic mesogenic epoxy resins.

This composite can be bonded to an alumina substrate with a copper foil on the other side, i.e., to a metal foil–laminated ceramic substrate. Table 2.8 shows examples of such characteristics [18]. Similar substrates for LED lighting devices have been developed. They are fabricated from crystalline epoxy resins obtained by introducing a mesogen skeleton [19]. Furthermore, other examples of high-thermal-conductivity glass composite substrate materials that have been investigated are new laminates such as a laminate of glass cloth base epoxy resin copper-clad laminate and a laminate of a non-woven glass fabric base epoxy resin copper-clad laminate (Table 2.9). The laminates filled with fillers of different sizes are fabricated. These composite substrates indicate a small drilling wear rate, high thermal conductivity, and good solder heat resistance. Such a substrate is employed as the LED mounting substrate.

Table 2.8 Representative characteristics of high-thermal-conductivity insulating sheet [18]

Properties	5W grade	10W grade (under development)
Thermal conductivity [W/(m·K)]	4.5–5.0	9.0–10.0
Glass transition temperature [°C]	170–180	190–200
Heat expansion coefficient (Linear expansion coefficient) α_1 [ppm/K]	25–30	20–25
Base resin decrease of 5% [°C]	290–300	290–300
Solder heat resistance	>280°C/5 min	280°C 5 min
Breakdown voltage [kV/200 μm]	5	4
Elastic modulus [MPa]	9–10	40
Shear bond strength [MPa]	6–8	4
Curing conditions	150°C/2 h + 180°C/2 h	140°C/2h + 190°C/2 h

Table 2.9 Characteristics of high-thermal-conductivity substrate [20].

Test items	Measurement method	Measurement conditions	Unit	Product [R-1787]	Comparative product 1 [CEM-3]	Comparative product 2 [FR-4]
Thermal conductivity	Laser flash	Ordinary	W/(m·K)	0.45	0.45	0.38
Insulating resistance	JIS C 6481	Ordinary	MW	5×10^8	5×10^8	5×10^8
Tracking resistance	IEC method	Ordinary	—	600	600	200
Dielectric constant	JIS C 6481	1 MHz	—	5.1	4.5	4.7
Dielectric loss tangent	JIS C 6481	1 MHz	—	0.016	0.015	0.015
Solder heat resistance	JIS C 6481	260°C	s	120 or more	120 or more	120 or more
T_g	TMA	—	°C	140	140	140
Thermal expansion coefficient (Thickness direction)	TMA	Room temperature to T_g	$10^{-6}/°C$	50	65	65

2.4.3.2 Solder resist films for a semiconductor package for solder resist films

A solder resist conductor package can be fabricated by covering a predetermined portion other than copper wiring region as shown in Fig. 2.48. For that purpose, the quality of insulation between wiring circuits and solders needs to be ensured even during solder reflow processes [21]. Solder resist materials are required to have a high T_g, a low thermal expansion coefficient, and excellent mechanical properties (such as heat cycle resistance). Furthermore, substrate materials with embedded lead wires and/or inter-terminal regions with a fine pitch also require ECM resistance. In order to obtain the performance as described above, an epoxy resin based, Cu-dispersed solder resist materials filled with nanofillers and micro filler have been developed [21]. It has been, furthermore, also reported that the EMC resistance of epoxy resins can be improved in EMC resistance by adding nanofillers [22].

Another report is that adding an elastomer such as butadiene rubber reduces stress and improves thermal shock resistance [23].

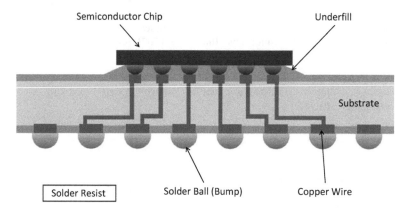

Figure 2.48 Semiconductor package substrate with flip-chip bonding [21].

Favorable data are obtained as follows: The resultant solder resist films have glass transition temperatures of 145–155°C and thermal expansion coefficients of 20–25 ppm/K. The synergy effects of nano- and microfillers are considered to be effective in improving the performance of materials used for electronics devices. Co-filling of both the fillers will provide a tool to minimize the size and the weight of the package and the assembly and achieve high density and high mounting, since various performance properties can be improved. Performance to be improved includes insulation degradation, dielectric constant, heat resistance, and thermal conductivity. It is hoped that nano-microcomposites will provide favorable solutions to accommodate the ever-increasing demands for further enhanced performance of materials to be used in the electronics sector.

References

1. Takagi, K. (2011). Until Capable of Wiring Board, *Nikkan Kogyo Shimbun*, Ltd., pp. 33 (in Japanese).
2. Honda, S. (2011). Electronic Component and Its Implementation, *Japan Institute of Electronics Packaging Spring Lecture Convention Materials*, pp. 513–516 (in Japanese).

3. Hozoji, H. (2012). Encapsulation Materials for Power Devices, *Journal of Japan Institute of Electronics Packaging*, **15**(5), pp. 374–378 (in Japanese).

4. Konda, S., Yoshimura, T., Saito, E., Hayashi, T., and Miwa, K. (2004). Nano-Hybrid Material-based Epoxy Resin for High Functionality, *Matsushita Electric Works Technical Report*, No. Feb. 2004, pp. 73–77 (in Japanese).

5. Ikezawa, R., Ishiguro, T., Hayashi, T., and Akagi, S. (2006). Flame Retardant-free System Halogen-free Sealing Material, *Hitachi Chemical Technical Report*, **46**, pp. 43–48 (in Japanese).

6. Asahi Denka Co., Ltd. and Kansai Electric Power Co., Ltd. (2005). Development of Nanotech Resin KA-100 to Withstand 400°C—To Open the Way for SiC Inverter Capacity-, *Asahi Denka News Release* (in Japanese).

7. Sugawara, Y., Miyanagi, Y., Asano, K., Agarwal, A., Ryu, S., Palmour, J., Shoji, Y., Okada, S., Ogata, S., and Izumi, T. (2006). 4.5 kV 120A SICGT and Its PWM Three Phase Inverter Operation of 100 kVA Class, *Proceedings of the 18th International Symposium on Power Semiconductor Devices & IC's*, pp. 117–120.

8. Nippon Shokubai (2012). SiC Power Applications Development Nanotechnology a Semiconductor for High-heat-resistant Sealant, *Nippon Shokubai News Release* (in Japanese).

9. Toray Dow Corning Co., Ltd., (2010). Joint Disclosure and NEDO in Next Generation Power Semiconductor for New Technology Development, Nano Tech 2010, *Toray News Release* (in Japanese).

10. Inada, T. (2009). Development of Die-Bonding Film for Semiconductor Packages by Applying Reaction-induced Phase Separation–Pursuing Soft, Endurable and Controllable Materials, *Hitachi Chemical Technical Report*, **52**, pp. 7–12 (in Japanese).

11. Gross, K., Hackett, S., Schultz, W., and Thompson, W. (2003). Nanocomposite Underfills for Flip-Chip Application, *2003 Electronic Components and Technology Conference*, p. 951.

12. Yamamoto, K., Shimada, Y., Hirata, Y., and Jindai, K. (2004). Trends in The Capacitor Built-in Substrate Materials, *Surface Technology*, **55**(12), p. 821.

13. Tsuyoshi, H., Asahi, N., Mizuguchi, T., and Nonaka, T. (2007). High Development of Dielectric Constant Photosensitive Polyimide, *17th Microelectronics Symposium (MES)*, p. 291.

14. KRI Corporation, (2006). Organic-inorganic Nanocomposite, *KRI Multi-Client Project Briefing Book* (in Japanese).

15. Makino, A., Arimura, M., Fujiyoshi, K., and Kuwabara, M. (2009). Print Particle Size on Particle Filling Rate of Development-BaTiO$_3$ Nanoparticles Deposited Thin Film Wiring High-Capacity Thin Film Capacitor for Board Built-in, The Effect of Dispersion of The Particles-, *Fukuoka Industrial Technology Center Research Report*, **19**, pp. 33–36 (in Japanese).

16. Arimura, M., Makino, A., Fujiyoshi, K., and Kuwabara, M. (2009). Development of A Printed Wiring High-capacity Thin Film Capacitor for Board Built-in - Preparation of Barium Titanate Nanoparticles Thin Film by Electrophoretic Deposition Method, *Fukuoka Industrial Technology Center Research Report*, **19**, pp. 29–32 (in Japanese).

17. Honda, S. (2012). *Yokohama Jisso Consortium 2012*, FIG. 88 (in Japanese).

18. Takezawa, Y. (2009). High Thermal Conductive Epoxy Resin Composites with Controlled Higher Order Structures, *Hitachi Chemical Technical Report*, **53**, pp. 5–16 (in Japanese).

19. Baba, D., and Sawada, T. (2011). Technological Trends of High Heat Dissipation Circuit Board Materials, *Panasonic Electric Works Technical Report*, **59**(1), pp. 17–23 (in Japanese).

20. Nozue, A., and Suzue, T. (2011). Glass-composite Circuit Board Material with High Thermal Conductivity, *Panasonic Electric Works Technique*, **59**(1), pp. 35–39 (in Japanese).

21. Yoshino, T., Joumen, M., and Katagi, H. (2006). Advanced Photo-Definable Solder Mask for High-performance Semiconductor Packages, *Hitachi Chemical Technical Report*, **46**, pp. 29–34 (in Japanese).

22. Ohki, Y., Horose, Y., Wada, G., Asakawa, H., Maeno, T., and Okamoto, K., (2012). Two Methods for Improving Electrochemical Migration Resistance of Printed Wiring Boards, *Proceedings of the 2012 International Conference on High Voltage Engineering and Application*, pp. 687–691.

23. Nagoshi, T., Tanaka, K., Yoshizako, K., Fukuzumi, S., and Kurabuchi, K. (2011). Semiconductor Package for Photosensitive Solder Resist Film FZ Series, *Hitachi Chemical Technical Report*, **54**, p. 30 (in Japanese).

24. TAIYO INK, PFR^{TM}-800 AUS^{TM}SR2-PKG for Development Type Solder Register Sulo Dry Film, *TAIYO INK MFG CO LTD News Release* (in Japanese).

Chapter 3

Compatibility of Dielectric Properties with Other Engineering Performances

Toshikatsu Tanaka[a], Toshio Shimizu[b], Muneaki Kurimoto[c], Yoshimichi Ohki[a], Norio Kurokawa[d] and Kenji Okamoto[e]

[a]*Waseda University*
[b]*Toshiba Corporation*
[c]*Nagoya University*
[d]*Japan Electrical Insulating and Advanced Performance Materials Industrial Association*
[e]*Fuji Electric Co., LTD.*

3.1 Composite Materials with Both Ample Thermal Conductivity and Voltage Endurance

Toshikatsu Tanaka

IPS Research Center, Waseda University, Kitakyushu-shi, Fukuoka 808-0135, Japan

t-tanaka@waseda.jp

Polymers naturally have low thermal conductivity. Polymer composites with practical high dielectric strength are required to have relatively high thermal conductivity. When polymers are loaded with high thermally conductive inorganic micro-fillers,

---

Advanced Nanodielectrics: Fundamentals and Applications
Edited by Toshikatsu Tanaka and Takahiro Imai
Copyright © 2017 Pan Stanford Publishing Pte. Ltd.
ISBN 978-981-4745-02-4 (Hardcover), 978-1-315-23074-0 (eBook)
www.panstanford.com

resulting composites have high thermal conductivity due to percolation effect of micro-fillers. However, they often suffer from reduced dielectric breakdown strength. In order to overcome this reduction, a certain nanotechnology method has been explored using nanofillers.

3.1.1 Thermal Conductivity Measured by Laser Flush Method

Thermal conductivity is usually measured by two types of measurement methods. Thermal conductivity is defined as the amount of heat flowing per unit area per unit time, when a plate with unit thickness has a unit temperature difference between its two opposite surfaces. The steady-state method and the non-steady-state laser flash method are often utilized. In the steady-state method, thermal conductivity is estimated from the temperature gradient when a sample is given steady heat flow in a one-dimensional axial direction or in the radial direction.

The laser flash method is suitable for thermal conductivity measurements of homogeneous materials, but is also adopted for composite materials. This method is briefly described here. Thermal conductivity is a product of thermal diffusivity and heat capacity. In this method, thermal diffusivity is obtained from heat response generated on one side surface when a sample is subjected to a laser light pulse on its other side surface. Heat capacity (a product of specific heat capacity and density of a sample) is independently obtained for the same sample by another method. Then, thermal conductivity is obtained from the following formula:

$$\lambda = \alpha C_p \rho \ [\text{W}/(\text{m·K})] \tag{3.1}$$

where λ, α, C_p, and ρ are thermal conductivity [W/(m·K)], thermal diffusivity [m^2/s], heat capacity [kg·K], and density [kg/m^3], respectively.

Values of thermal conductivity λ are 298 and 236 for copper and aluminum as metals, respectively, while they are 0.41, 0.21 and 0.16 for polyethylene, epoxy (bisphenol A type) and silicone rubbers as polymers, respectively. The thermal conductivities of polymers are less than 1/500 those of metals. It is reported that thermal conductivity is 1.5 to 1.6, 36 to 42, and 150 to 220 for silica, alumina, boron nitride (BN), and aluminum nitride (AlN) as inorganic substances, respectively.

A laser flash measurement system is schematically depicted in Fig. 3.1a [1]. A flat specimen with thickness l is set in a vacuum heating oven. One surface of the specimen is heated by laser pulse light irradiation, and temperature response appearing on its opposite surface is measured. Temperature response with time is shown in Fig. 3.1b [1]. Based on the half-time method, thermal diffusivity α can be obtained from the following formula (3.2) using half of the time for temperature saturation $t_{1/2}$.

$$\alpha = 0.139 \frac{l^2}{t_{1/2}} \quad (3.2)$$

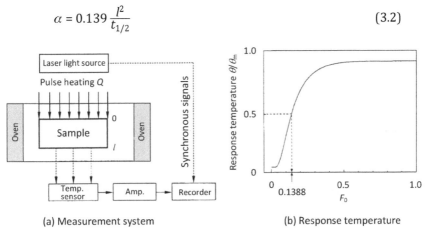

(a) Measurement system (b) Response temperature

Figure 3.1 Laser flash method for heat diffusivity measurement and half time calculation method [1].

3.1.2 Thermal Conductivity Increased by Loading Micro-Filler into Polymers

Thermal conductivity is usually as low as the order of 10^{-1} W/(m·K). Composite materials with relatively high thermal conductivity can be obtained by adding ceramic particles with high thermal conductivity ranging from 10 to 100 W/(m·K) to polymers. Several formulae to estimate thermal conductivity are proposed for composite systems where particles are non-oriented or oriented in their dispersion phase. Bruggeman's formula is often used to calculate a theoretical reference value. Thermal conductivity is obtained first for a composite system including one spherical filler to make a unit cell, and then for a continuous set of unit cells. The resulting formula is given as follows [2]:

$$V_f = 1 - \frac{(\lambda f - \lambda c)}{(\lambda f - \lambda p)}\left(\frac{\lambda p}{\lambda c}\right)^{1/3} \qquad (3.3)$$

where suffixes p, f, and c represent polymer, filler, and composite, respectively. Filler volume is given as a function of thermal conductivity of composite components. Thermal conductivities of composites are not presented explicitly, but can be calculated as a function of the volume fraction of filler particles, as shown in Fig. 3.2.

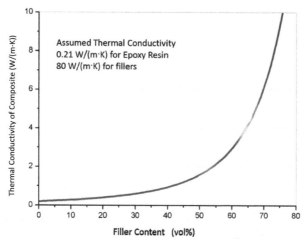

Figure 3.2 Thermal conductivity of an epoxy/filler composite calculated based on Bruggeman's formula.

Figure 3.2 represents the thermal conductivity of composites as a function of a loading fraction of filler particles as obtained from the formula (3.3), where thermal conductivity is 0.12 and 80 W/(m·K) for polymer (bisphenol-A epoxy) and filler particles, respectively. It is indicated that the thermal conductivity increases sharply in the region of large filler loadings. This is considered to take place due to percolation in which filler particles are physically overlapped beyond 60 vol% loadings.

Some of the thermal conductivity values experimentally obtained are listed in Table 3.1 [3, 4]. The highest value in this table is 32.5 W/(m·K) for a composite of BN and polybenzoxazine resins. These results demonstrate that high-thermal-conductivity composite materials can be fabricated for practical use. For example, thermal conductivity is required to be higher than 10 W/(m·K) for PWB (printed wiring boards) used in microelectronics. This kind of composite substance is considered to be in a feasibility range for future practical use. In the electric power sector, nanocomposites are preferred since they have high withstand electric strength. Furthermore, if nanocomposites can achieve 1 W/(m·K) of thermal conductivity, they enable greatly improved electric rotating machine insulation.

Table 3.1 Examples of high-thermal-conductivity polymer composites

Fillers	Polymer matrices	Filler size (μm)	Filler content (%)	Thermal conductivity (W/(m·K))	CA/ST
BN	Polybenzoxazine	225	78.5 vol.	32.5	no CA
BN-MM	Epoxy	0.6 (h), 1(c)	27 vol.	19.0	no CA
BN	Epoxy	5–11	57 vol.	10.3	silane
BN	PI	8	60 wt.	7	—
BN	Epoxy	5–11	57 vol.	5.3	no
BN	Epoxy	CGM	64.9 wt.	5.13	no ST
BN-nT	PS	50 nmφ/4–10	35 wt.	3.61	no ST
BN	Epoxy	50–100 CGM	80 wt.	3.5	—
BN	Mesogen epoxy	5.5	35 vol.	2.2	no ST
BN-MM	PCB-filled epoxy	4, 0.15, 0.053	30 vol.	1.5	silane
BN	PBT	2–3, 5–11 CGM	20 wt.	1.18	no ST
BN-n	Epoxy	53 nm	30 wt.	0.9	silane
BN-n	Epoxy	—	37wt.	0.7	ST
BN	Silica Filled epoxy	9–12	21.7 wt.	0.59	—
BN	Epoxy	9–12	20 wt.	0.54	—
BN	FRP-filled epoxy	9–12	—	0.54	no silane
BN-nT	PVA	57 nmφ/7–11	3 wt.	0.3	Catechin
BN-n	Epoxy	8–10 layers of 2.5–3 nm	10 wt.	0.20	silane

(*Continued*)

Table 3.1 (*Continued*)

Fillers	Polymer matrices	Filler size (μm)	Filler content (%)	Thermal conductivity (W/(m·K))	CA/ST
AlN	Epoxy	7	60 vol.	11.0	silane
AlN	PVDF	12 Whiskers	60 vol.	7.4	no ST
AlN-n	Epoxy	0.5 Whiskers	47 vol.	4.2	no ST
AlN	Brominated EP	2.3	40 wt.	1.0	silane
AlN-n	Epoxy	20–500 nm	10 wt.	0.2	silane
Al_2O_3	Mesogen epoxy	—	—	10	—
				5	—
Al_2O_3	Epoxy	10	55 vol.	5	no ST
Al_2O_3	Epoxy	<25.8	75	1.29	no ST
Al_2O_3	Epoxy	4–20	60 wt.	0.68	—
SiO_2	Epoxy	4–20	60 wt.	0.75	—
SiO_2	Epoxy	<50	65	0.7	no ST
SiC	Epoxy	500 nm	27 wt.	0.32	—
AlN/BN Hybrid	PI	79/—	70 vol.	9.3	—
SiC/CNT Hybrid	Epoxy	SiC: 50 nm 40–80φ × 5–15	30 vol.	2.1	silane no ST
Diamond	Epoxy	<10	70 vol.	4.1	—
Diamond	Silica filled epoxy	3–6	7.29 wt	0.92	—
Ag-nW	Silicone	0.1φ × 5–50	7.2 wt.	0.19	no ST
No Filler	Lyotropic LC-PBO	—	—	20	—

Abbreviations: MM: multi-modal particle size mixing; Hybrid: hybrid mixing; CA: coupling agents; ST: surface treatment; CGM: conglomerated.

3.1.3 Further Increase in Thermal Conductivity Is Realized by Improvement of Interfaces

Thermal conductivity is governed by phonon flow. When interfaces are discontinuous, the thermal flow is hindered. Therefore, thermal conductivity can be furthermore increased, if the interfaces are modified to have as much continuity as possible.

A trial was made to increase thermal conductivity by loading AlN (aluminum nitride) fillers into epoxy resins. Nanocomposites were fabricated using several kinds of fillers: three kinds of silane coupling agents (end groups with epoxy, mercapto and amino), POSS (polyhedral oligomeric silsesquioxane), hyper-branched polymers, and oxidized graphene, and their thermal conductivity was evaluated [5]. One of the results is shown in Fig. 3.3. A bisphenol-A epoxy resin (thermal conductivity: 0.21 W/(m·K)) are loaded with AlN fillers (size: 1.1 μm and thermal conductivity: 150 to 220 W/(m·K)) to fabricate high-thermal-conductivity composites. Silane in this case has a mercapto functional group as its end group. It was found that thermal conductivity increases up to 10 times if the epoxy resin is loaded with AlN fillers by 50 vol%. Thermal conductivity tends to saturate above this volume fraction. It will increase further up to 65 vol% if surface treatment is carried out. Above this loading fraction, it does not increase with the present technology.

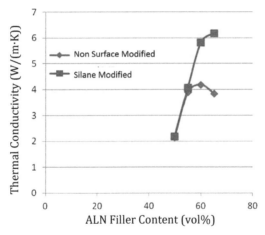

Figure 3.3 Dependence of thermal conductivity of epoxy/AlN microcomposite on filler content.

Increase in thermal conductivity of composites by infilling high-thermal-conductivity fillers is governed by (i) propagation speed of phonon and (ii) percolation (contact between neighboring filler particles). It is well recognized that the former is predominant in low loading content, while the latter is more effective in high loading content. In general, it is difficult to increase thermal conductivity in high loadings as compared to theoretical expectation. This is due to technical problems such as the formation of internal voids caused by the increase in viscosity of the composites. It is necessary to develop new methods to fabricate composites with low void content for high filler loadings.

3.1.4 Voltage Endurance Decreased by Micro-Filler Loading

Generally dielectric breakdown voltage decreases when epoxy resins are loaded with micro-fillers. Since this reduction is not desirable, it should be mitigated. Figure 3.4 shows the dependence of dielectric breakdown strength on loading content of micro-fillers in the case of spherical alumina-filled epoxy resins [6]. It is clear from this figure that dielectric breakdown strength decreases from 200 to 60 kV$_{rms}$/mm as filler loading increases from 0 to 80 vol%. Such a big decrease is considered to originate from the increased void formation rate with the increase in filler loadings.

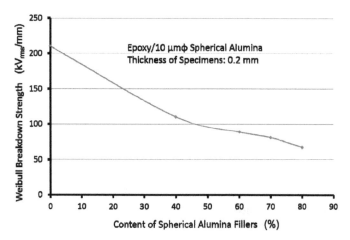

Figure 3.4 Dependence of breakdown strength of epoxy/AlN micro-composite on filler content (an example).

In the case of low loading fraction below 10 vol%, dielectric breakdown strength does not necessarily decline. It depends on the kind of fillers and loading method. Surface treatment or silane coupling around the surfaces of fillers can increase the threshold loading content at which dielectric breakdown strength starts to decline. Plate-like fillers can suppress this declining tendency. For example, epoxy resins with BN fillers (size: 5 µm, content: 10 vol%) exhibit more or less higher DC dielectric breakdown strength than neat epoxy. Plate-like particles such as BN fillers are horizontally oriented when composites are hot-press-molded, and then such composites are more resistant to longitudinal electric field stress.

3.1.5 Recipe for Compatibility: An Exquisite Combination of Nano- and Microfillers

It was found that polymers exhibit lower dielectric breakdown strength if they are filled with large amounts of micro-fillers to enhance their thermal conductivity. It is necessary to achieve high thermal conductivity and high dielectric breakdown strength simultaneously for practical use. For that purpose, a method of nano- and microfiller co-mixing was explored. Figure 3.5 shows a conceptual figure for comparison of dielectric breakdown strength between micro- and nanocomposites. A 0.2 mm-thick model board for PWB is assumed. Withstanding voltage required for this board is 5 kV_{rms}. Its reference line is indicated on the vertical axis of the figure. When an epoxy resin is filled with a large amount of BN, AlN or Al_2O_3 fillers with high thermal conductivity, the resulting composite obtains a target value of high thermal conductivity such as 10 W/(m·K). However, it suffers from a reduction in withstanding voltage much lower than the target value. This once-lowered value is expected to recover to a certain higher value as indicated by an arrow in the figure if nanofillers are added to the epoxy resin.

Three kinds of specimens, a base epoxy resin, a micro-composite, and a nanomicrocomposite, are evaluated based on surface erosion caused by application of AC voltage (4.8 kV_{rms} for 2 h.) using an electrode system as indicated in the upper part of Fig. 3.6. Some results are shown for erosion and breakdown in Fig. 3.6. The micro-composite suffers from dielectric breakdown due to deep erosion. The nanofiller-added microcomposite remains almost undamaged. Dielectric breakdown time until breakdown that takes place under a constant voltage such as 4 kV_{rms} is also evaluated. This value reduces by half for the micro-composite but enhances to 1.5 times for the nanomicrocomposites compared with that of the base resin. This demonstrates the excellence of nanofiller addition [7].

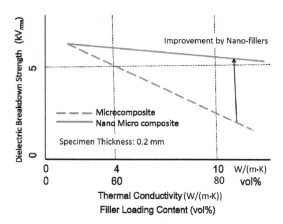

Figure 3.5 Comparison of dielectric breakdown voltage between micro-composites and nanomicrocomposites.

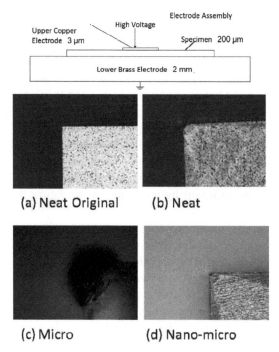

Figure 3.6 Comparison of dielectric breakdown voltage between micro-composites and nanomicrocomposites. © 2010 IEEE. Reprinted, with permission, from Li, Z., Okamoto, K., Ohki, Y., and Tanaka, T. (2010). Effects of Nanofiller Addition on Partial Discharge Resistance and Dielectric Breakdown Strength of Micro-Al_2O_3/Epoxy Composite, *IEEE Trans. Dielectr. Electr. Insul.*, **17**(3), pp. 653–661.

A mechanism is proposed to explain this phenomenon as shown in Fig. 3.7 [8]. The breakdown strength is reduced due to the defects or voids possibly formed around micro-fillers in micro-composites. Nanofillers, when added, work effectively to suppress breakdown propagation by the following possible processes:

(i) Breakdown propagation is enormously retarded in nanofiller regions.

(ii) Breakdown propagation is retarded due to nanofillers existing in interfaces between epoxy resins and microfillers.

It is necessary to develop superb nanomicrocomposites with both high thermal conductivity and high dielectric breakdown strength utilizing the natural properties of nanofillers. However, it is also necessary to pay attention to the following problems. Enhancement of breakdown strength due to nanofiller addition is sometimes less than expected. Breakdown strength is reduced from 200 kV$_{rms}$/mm to 90 kV$_{rms}$/mm for 0.2 mm thick specimens when a base epoxy resin is loaded with alumina fillers (size: 10 μm, loading content: 60 wt%). Addition of nanoalumina fillers (size: 7 nm, loading content to epoxy resin: 5 wt%) improves breakdown strength by only a few %. Appropriate methods of fabricating nanomicrocomposites should be further explored, focusing on reduction of voids and/or defects between resins and micro-fillers, as described in Fig. 3.7.

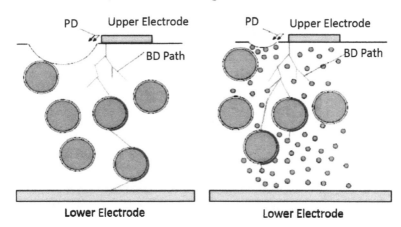

Large Circles: micro-fillers, Small Dots: nano-fillers

Figure 3.7 Schematics of dielectric breakdown propagation processes proposed for micro-composites and nanomicrocomposites [8].

3.1.6 Composite with High Thermal Conductivity and Sufficient Voltage Endurance

Values of thermal conductivity such as 5 to 10 W/(m·K) can be easily obtained when polymers are modified to composite substances. But it is necessary to develop some methodology in order to maintain sufficient dielectric breakdown strength. Figure 3.8 shows characteristics of a composite fabricated to possess both natural performances. An epoxy resin loaded with micro (BN + AlN) fillers by 65 vol% achieves thermal conductivity as high as 12.3 W/(m·K), and at the same time exhibits withstanding voltage exceeding the target voltage such as 5 kV$_{rms}$ (withstanding stress: 20 kV$_{rms}$/mm) [9, 10]. It was clarified that BN fillers do not cause much reduction in breakdown strength. Nevertheless, the addition of nanofillers enhances breakdown strength further, say, by 20%. As such, appropriate methods for improvement of polymer characteristics by the combination of nano- and microfillers must be a big step in technological innovation and evolution in the field of dielectrics and insulating materials.

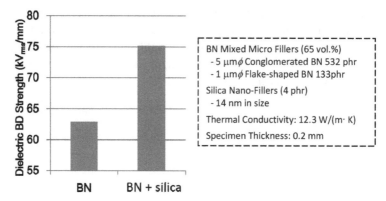

Figure 3.8 Improvement of dielectric breakdown strength of high thermal conductivity micro-composite by adding nanofillers.

References

1. Hatta, I., ed. (2003). *Cutting-Edge Thermal Measurement Methods: From Fundamentals to Applications* (Agune Technology Center, in Japanese).

2. Wong, C. P., and Bollampally, R. S. (1999). Thermal Conductivity, Elastic Modulus, and Coefficient of Thermal Expansion of Polymer Composites Filled with Ceramic Particles for Electronic Packaging, *J. Appl. Polym. Sci.*, **74**, pp. 3396–3403.

3. Huang, X., Jiang, P., and Tanaka, T. (2011). A Review of Dielectric Polymer Composites with High Thermal Conductivity, *IEEE EI Mag.*, **27**(4), pp. 8–16.

4. Tanaka, T., Kozako, M., and Okamoto, K. (2011). Toward High Thermal Conductivity Nano Micro Epoxy Composites with Sufficient Endurance Voltage, *Proceedings of the IEEE International Conference on Electrical Engineering (ICEE)*, No. D7-A032, p. 6.

5. Huang, X., Iizuka, T., Jiang, P., Ohki, Y., and Tanaka, T. (2012). The Role of Interface on the Thermal Conductivity of Highly Filled Dielectric Epoxy/AlN Composites, *J. Phys. Chem. C*, **116**, pp. 13629–13639.

6. Li, Z., Okamoto, K., Ohki, Y., and Tanaka, T. (2011). The Role of Nano and Micro Particles on Partial Discharge and Breakdown Strength in Epoxy Composites, *IEEE Trans. Dielectr. Electr. Insul.*, **18**(3), pp. 675–681.

7. Andritsch, T., Kochetov, R., Gebrekiros, Y. T., Morshuis P. H. F., and Smit, J. J. (2010). Short Term DC Breakdown Strength in Epoxy Based BN Nano- and Microcomposites, *Proc. IEEE ICSD*, No. B2-3, pp. 179–182.

8. Li, Z., Okamoto, K., Ohki, Y., and Tanaka, T. (2010). Effects of Nanofiller Addition on Partial Discharge Resistance and Dielectric Breakdown Strength of Micro-Al_2O_3/Epoxy Composite, *IEEE Trans. Dielectr. Electr. Insul.*, **17**(3), pp. 653–661.

9. Wang, Z., Iizuka, T., Kozako, M., Ohki, Y., and Tanaka, T. (2011). Development of Epoxy/BN Composites with High Thermal Conductivity and Sufficient Dielectric Breakdown Strength Part I. Sample preparation and Thermal conductivity, *IEEE Trans. Dielectr. Electr. Insul.*, **18**(6), pp. 1963–1972.

10. Wang, Z., Iizuka, T., Kozako, M., Ohki, Y., and Tanaka, T. (2011). Development of Epoxy/BN Composites with High Thermal Conductivity and Sufficient Dielectric Breakdown Strength Part II. Breakdown Strength, *IEEE Trans. Dielectr. Electr. Insul.*, **18**(6), pp. 1973–1983.

3.2 Composites with Both Low Thermal Expansion Coefficient and High Voltage Endurance

Toshio Shimizu[a] and Muneaki Kurimoto[b]

[a]*Power and Industrial Systems R&D Center,*
Toshiba Corporation, Fuchu-shi, Tokyo 183-8511, Japan
[b]*Department of Electrical Engineering and Computer Science,*
Nagoya University, Nagoya-shi, Aichi 464-8603, Japan

toshio4.shimizu@toshiba.co.jp, kurimoto@nuee.nagoya-u.ac.jp

It is possible to improve electric stress resistance and decrease thermal expansion coefficient together by applying nanocomposites. Furthermore, thermal expansion coefficient can be decreased up to the level of metals by filling with both nanofiller and microfiller. Nanomicro hybrid materials are expected to be applied to polymeric molded products for high-voltage apparatuses.

3.2.1 Thermal Expansion Coefficient Is an Important Material Property of Molded Products

The length and volume of materials change with temperature as they expand or shrink. The rate of this change with unit temperature change is the thermal expansion coefficient. If one of the dimensions of the material is L and the temperature increase is ΔT, then the length change ΔL is given by

$$\Delta L = \alpha L \Delta T \tag{3.4}$$

where α is the coefficient of linear thermal expansion.

The coefficients of linear thermal expansion of solid materials are measured by thermal mechanical analysis (TMA) and they vary according to material composition. The coefficients of thermal expansion of typical electrical materials are summarized in Table 3.2. Conventional insulating molded products used in power apparatuses have the structure of conductors covered with insulation materials. The coefficients of linear thermal expansion of insulating polymers, such as epoxy resin are larger than those of conductive metals, such as aluminum or copper. When materials with different coefficients of thermal expansion are combined, large mechanical stresses are generated because of their different levels of expansion or shrinkage with temperature change. This causes problems such as cracking or debonding at interfaces.

Table 3.2 Coefficients of linear thermal expansion of electrical materials

Material	Coefficient of linear thermal expansion [$\times 10^{-6}\,K^{-1}$]
Steel	12.1
Aluminum	23.1
Copper	16.8
Gold	14.3
Silica (crystal)	0.56
Alumina	7
Epoxy resin	62
Polypropylene	110

Consider a structure 1 m long comprising steel bonded to epoxy resin at 20°C. If this structure is heated uniformly to 70°C, the length change ΔL of the epoxy resin will be 3.1 mm, while that of the steel will be 0.605 mm. This will create shear stress at the adhesive interface between the two materials. This stress is called thermal stress. Because various materials are used in combination in electrical apparatuses, it is necessary to consider thermal stress.

The cross section of a tank in a gas-insulated switchgear (GIS) is shown in Fig. 3.9. Some epoxy molded parts called "spacer" are used to separate a gas insulated line. This structure consists of an aluminum bus bar molded with epoxy resin. It is important to attain strong adhesion between the aluminum bus bar and the epoxy resin to prevent leakage of SF_6 gas and to support the bus bar in a predetermined position. A large current flows during bus bar in operation, so the temperature rises due to Joule heat. Because the epoxy resin's expansion with temperature rise is larger than that of aluminum, thermal stress occurs at the adhesion interface of the aluminum and epoxy resin, causing thermal stress. This can lead to debonding or cracking at the adhesion interface, resulting in damage to the apparatus. Therefore, the designer must try to lower the difference between the thermal expansion coefficients of these materials. A large amount of ceramic filler, such as alumina or silica, may be added to the epoxy resin to decrease its thermal expansion coefficient.

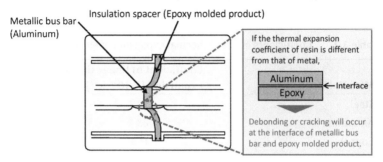

Figure 3.9 Cross section of GIS insulation spacer.

In this way, a variety of molded parts using a polymer in combination with metal and ceramics are applied to insulation structures such as power apparatuses and electronic parts. It is important to improve the electric stress resistance of these polymers and to reduce their linear coefficient of thermal expansion.

3.2.2 It Is Possible to Decrease the Thermal Expansion Coefficient and to Improve the Electric Stress Resistance Together by Applying Nanocomposites

Expansion or shrinkage of an object with temperature depends on increasing or decreasing its intermolecular or interatomic distance. Because restriction between molecules is particularly low in polymers, they have larger thermal expansion coefficients than other materials. It was reported that the thermal expansion coefficients of thermoplastic polymers are reduced by adding exfoliated organic-modified clay and dispersing them in nano-order uniformly in small quantities. It is considered that nanoparticles restrict the thermal movement of molecules.

In contrast, for thermosetting resins such as epoxy resin, a low thermal expansion coefficient is accomplished by adding clay to be the nanocomposite [1]. As an example, the linear coefficient of thermal expansion and the glass transition temperature (T_g)

of an epoxy resin filled with triamine compound ornamentation clay are shown in Table 3.3. The linear expansion coefficient can be reduced by kneading with a roller, exfoliating the clay, and raising the dispersion degree of nanoparticles. However, because triamine compound reacts with epoxy resin, vibrates itself and reduces the cross-linked density between clay layers, glass transition temperature and thermal resistance decrease. In order to attain a lower thermal expansion coefficient for epoxy resin by forming nanocomposite, it is important to choose an appropriate organic modifier, and to suppress the drop of glass transition temperature.

Table 3.3   Effect of triamine compound ornamentation clay [1]

Clay (filling amount: 8.4vol%)	Coefficient of linear thermal expansion [× 10^{-6} K^{-1}]	Glass transition temperature [°C]
Non-treated	63.1	179.7
A-BHMT* treated	55.4	184.4
A-BHMT treated + roll mixing	52.6	171.4

*Acetyl bis(hexamethylene)triamine.

A needle electrode 1 mm in diameter, tip angle of 30°, tip radius of 5 μm and a gap of 3 mm was molded with epoxy resin as shown in Fig. 3.10. A voltage impression test was carried out at AC 10 kV and 1 kHz with the needle–plate electrode, and the time to break down was measured. When the temperature increased to 145°C, around glass transition temperature, the time to break-down increased. For the nanocomposite of this epoxy resin filled with 5 wt% of clay organic modified with quaternary alkyl ammonium, the time to breakdown increased over the whole temperature range and the electric stress resistance was improved.

The residual stress around the molded needle electrode was measured by photoelasticity, as shown in Fig. 3.11. The stress was larger when the density of the obis like straps increased, and it increased with decreasing temperature. The residual stress was decreased by applying nanocomposite. This is caused by lower thermal expansion and the lower hardening shrinkage of nanocomposites, and the decreasing residual stress contributes to the improvement of electrical stress resistance.

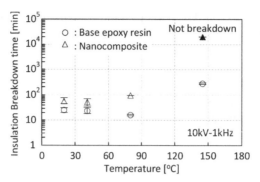

Figure 3.10 Temperature dependence of insulation breakdown time under constant AC voltage (10 kV^{-1} kHz) (a filled symbol indicates no breakdown). © 2006 IEEE. Reprinted, with permission, from Imai, T., Sawa, F., Ozaki, T., Shimizu, T., Kido, R., Kozako, M., and Tanaka, T. (2006). Influence of Temperature on Mechanical and Insulation Properties of Epoxy-Layered Silicate Nanocomposites, *IEEE Trans. Dielectr. Electr. Insul.*, **13**(1), pp. 445–452.

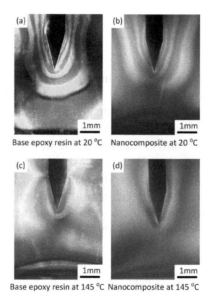

Figure 3.11 Internal stress observed by a polarizing microscope. © 2006 IEEE. Reprinted, with permission, from Imai, T., Sawa, F., Ozaki, T., Shimizu, T., Kido, R., Kozako, M., and Tanaka, T. (2006). Influence of Temperature on Mechanical and Insulation Properties of Epoxy-Layered Silicate Nanocomposites, *IEEE Trans. Dielectr. Electr. Insul.*, **13**(1), pp. 445–452.

3.2.3 Hybrid Filling with Nano- and Microfillers Realizes Further Low Thermal Expansion and Drastically Improves Electrical Stress Resistance

Although the coefficient of thermal expansion of polymers is decreased by adding nano-clay, it is not sufficient to combine them with metals or ceramics. Therefore, combination of a large amount of conventional micro-filler and a small amount of nanofiller is being studied. Table 3.4 shows the linear coefficient of thermal expansion coefficient of a conventional casting material, which has the basic composition of epoxy resin and acid anhydride hardener filled with a large amount of silica filler. The linear thermal expansion coefficient of the material filled with a small amount of organic-modified clay to improve compatibility with epoxy resin is also shown. Generally, the average dimension of silica fillers is micro-order. In the examples in Table 3.4, composition of the conventional materials is controlled to decrease the linear thermal expansion coefficient to that of aluminum ($23.1 \times 10^{-6} K^{-1}$), and this is determined by the volume fraction of micro-filler.

Table 3.4 Components for conventional filled epoxy resin and NMMC

Specimen		Conventional filled epoxy	NMMC (OMLS 0.3 vol%)	NMMC (OMLS 1.5 vol%)
Composition [phr]	Epoxy resin	100	100	100
	Hardener	86	86	86
	OMLS	0	2.3	9.8
	Silica filler	340	340	340
Coefficient of linear thermal expansion [$\times 10^{-6} K^{-1}$]		23.8	—	24.0

Source: © 2006 IEEE. Reprinted, with permission, from Imai, T., Sawa, F., Nakano, T., Ozaki, T., Shimizu, Kozako, M., and Tanaka, T. (2006). Effect of Nano- and Microfiller Mixture on Electrical Insulation Properties of Epoxy Based Composites, *IEEE Trans. Dielectr. Electr. Insul.*, **13**(1), pp. 319–326.
Note: Phr: parts per hundred parts of resin by weight; OMLS: organic modified layered silicate.

In this case, the decrease of the linear coefficient of thermal expansion is dominated by a large amount of silica filler, and the effect of filling with some percentage of nanofiller is very small. Adding a large amount of nanofiller according to its volume fraction is expected to decrease the linear coefficient of thermal expansion, while the viscosity increases dramatically and there is a limit to nanofilling.

On the other hand, Weibull plots of insulation breakdown strength obtained with the same materials by a needle-plate electrode are shown in Fig. 3.12. Specimens of epoxy resin containing layered silicate and micro-scale silica fillers (Nano- and microfiller mixture composite: NMMC) are expected to improve insulation breakdown strength. Especially, a conventional material filled with 1.5 vol% of nanofiller showed a large increase in insulation breakdown strength.

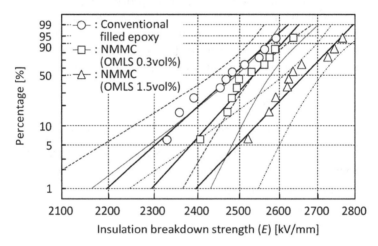

Figure 3.12 Weibull plots of insulation breakdown strength (confidence intervals of Weibull plots are relevant to probability 95%). © 2006 IEEE. Reprinted, with permission, from Imai, T., Sawa, F., Nakano, T., Ozaki, T., Shimizu, T., Kozako, M., and Tanaka, T. (2006). Effect of Nano- and Microfiller Mixture on Electrical Insulation Properties of Epoxy Based Composites, *IEEE Trans. Dielectr. Electr. Insul.*, **13**(1), pp. 319–326.

Figure 3.13 shows the insulation breakdown time obtained by a needle-plate electrode under constant AC voltage. Compared with the short time insulation breakdown strength, the effect of nanofiller is obvious in low electric fields and longer life can be expected. It was confirmed that nano-silica and nano-titan oxide showed the same effects as clay.

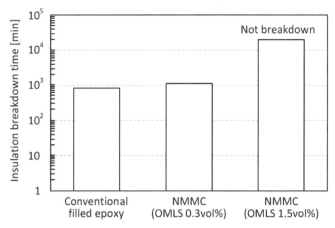

Figure 3.13 Comparison of insulation breakdown time under constant AC voltage (10 kV^{-1} kHz). © 2006 IEEE. Reprinted, with permission, from Imai, T., Sawa, F., Nakano, T., Ozaki, T., Shimizu, Kozako, M., and Tanaka, T. (2006). Effect of Nano- and Microfiller Mixture on Electrical Insulation Properties of Epoxy Based Composites, *IEEE Trans. Dielectr. Electr. Insul.*, **13**(1), pp. 319–326.

The estimated model for the mechanism of this improvement on insulation properties is shown in Fig. 3.14. Even epoxy resin filled with micro-silica has a longer breakdown life than neat epoxy resin. Electrical treeing is formed around the edge of the needle electrode with a high electrical field, which propagates to the counter electrode and breaks down when reaching a certain length. The damaged component is organic resin so that the treeing propagates in the resin. If inorganic silica fillers with higher insulation strength are dispersed uniformly and adhere to the resin tightly, the propagation of electrical treeing is disturbed and breaks into branches, as shown in Fig. 3.14. As a result, propagation is delayed and breakdown life increases.

Furthermore, branching of this electrical treeing is accelerated by filling with nanofiller closely in resin between micro-silicas, and more energy is required to propagate electrical treeing. Even a small amount of nanofiller has the advantage of decreasing the distance between fillers.

As stated above, molded products with a combination of polymers and metals or ceramics form higher thermal stress with temperature change, causing cracks or debonding. To prevent this, it is important to decrease the coefficient of thermal expansion of polymers. Although microfillers are more effective in decreasing the coefficient of thermal expansion of polymers, further addition of nanofiller could dramatically improve electrical insulation properties.

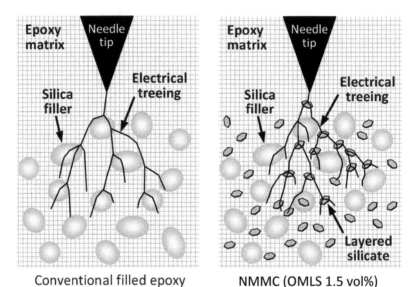

Figure 3.14 Estimated model for effect of layered silicate and silica filler mixture (nano- and microfiller mixture). © 2006 IEEE. Reprinted, with permission, from Imai, T., Sawa, F., Nakano, T., Ozaki, T., Shimizu, Kozako, M., and Tanaka, T. (2006). Effect of Nano- and Microfiller Mixture on Electrical Insulation Properties of Epoxy Based Composites, *IEEE Trans. Dielectr. Electr. Insul.*, **13**(1), pp. 319–326.

References

1. Konda, T., Yoshimura, T., Saito, E., Hayashi, T., and Miwa, K. (2003). Low Thermal Expansion Nanocomposites Consisting of an Epoxy Resin and Clay with Organic Modifications, *J. Networkpolymer.*, **24**, pp. 46–52 (in Japanese).

2. Imai, T., Sawa, F., Ozaki, T., Shimizu, T., Kido, R., Kozako, M., and Tanaka, T. (2006). Influence of Temperature on Mechanical and Insulation Properties of Epoxy-Layered Silicate Nanocomposites, *IEEE Trans. Dielectr. Electr. Insul.*, **13**(1), pp. 445–452.

3. Imai, T., Sawa, F., Nakano, T., Ozaki, T., Shimizu, Kozako, M., and Tanaka, T. (2006). Effect of Nano- and Microfiller Mixture on Electrical Insulation Properties of Epoxy Based Composites, *IEEE Trans. Dielectr. Electr. Insul.*, **13**(1), pp. 319–326.

4. Imai, T., Komiya, G., Murayama, K., Ozaki, T., Sawa, F., Shimizu, T., Harada, M., Ochi, M., Ohki, Y., and Tanaka, T. (2008). Improving Epoxy-Based Insulating Materials with Nano-Fillers Toward Practical Application, *IEEE Int. Symp. Electr. Insul.*, No. S3-1, pp. 201–204.

3.3 Composites with High Permeability and High Dielectric Permittivity

Yoshimichi Ohki

*Department of Electrical Engineering and Bioscience
and Research Institute of Materials Science and Technology,
Waseda University, Shinjuku-ku, Tokyo 169-8555, Japan*

yohki@waseda.jp

Various trials to make magneto-dielectric nanocomposites have been carried out to develop a new substrate that can exhibit both good electrical insulating characteristics with high permittivity and good magnetic characteristics with high permeability. An electromagnetic wave absorber to minimize electromagnetic interferences and a new substrate capable of miniaturizing antennas in digital broadcast systems are their potential applications.

3.3.1 For What Are Magnetized Dielectrics Used?

In July 2011, all the surface television broadcasts in Japan, via both the surface and satellite, were switched to a digital system. In response, installation of digital broadcast receivers in portable information devices has been accelerating. Since the lowest frequency of the digital surface television broadcasts in Japan is 470 MHz, its wavelength λ is about 640 mm. Therefore, the necessary length of the quarter-wavelength antenna is 160 mm, which is longer than half the length of ordinary cellular phones. As λ depends on the dielectric permittivity (ε) and magnetic permeability (μ) of the insulating substrate around the antenna in the form of $\lambda \propto (\varepsilon\mu)^{-1/2}$, materials with high ε values have been mainly used to make antennas compact.

However, recently, magneto-dielectric composites with high μ are gaining attention [1–6]. Ferrite, which is widely used as a magnetic material, exhibits a low eddy current loss, since its electrical conductivity σ is low and μ is also low at GHz frequencies because of its small magnetic moment [3]. On the other hand, mass or bulk iron (Fe) exhibits large eddy current losses and cannot serve as an insulating substrate because of high σ, while it has a large magnetic moment. Therefore, as a very unique example of making polymer nanocomposites, trials to develop an electrical insulating polymer that possesses both high dielectric permittivity and high magnetic permeability have been carried out.

Compared to the above-mentioned objective, another purpose of making such composites is pure scientific. A typical case is a study on the minimum size necessary for ferromagnetic materials to exhibit ferromagnetic multi-domain structures. It has also been reported that ferromagnetic nanoparticles smaller than a critical size possess a single domain structure, leading to magnetic properties superior to those of bulk materials.

3.3.2 What Kind of Magnetized Dielectrics Are Available?

The magneto-dielectric composites that have been examined so far can be categorized by their materials into the following three categories: the addition of (ferro)magnetic materials into dielectric substances such as ceramics, that of magnetic materials into polymers, and that of both ferroelectric and ferromagnetic materials into insulating substances. Various ceramics such as TiO_2, SiO_2, and B_2O_3, and various polymers such as ethylene propylene diene copolymer and low-density polyethylene have been tried, while Co-Pt, Fe-Pt, Co, Fe_3O_4, Fe, and $CoFe_2O_4$ have been tried as ferromagnetic materials and TiO_2 has been used as a high-permittivity materials [1–8].

For example, for the purpose of studying the effect of using magneto-dielectric nanocomposite materials on the miniaturization of the physical dimension of an active antenna, K. Borah and N. S. Bhattacharyya of Microwave Research Laboratory in India reported that the size of a microstrip antenna can be miniaturized if nano-sized $CoFe_2O_4$ is incorporated into low-density polyethylene using a co-precipitation process after nanosized ferrite had been obtained by a heat treatment of precipitate in a nitrate precursor [3].

A US researcher and his co-workers made composites of Fe_3O_4 nanoparticles and block copolymer of styrene-b-ethylene/butylene-b-styrene (SEBS) [8] and found that the bandwidth for antenna applications can be improved significantly at a proper ratio of dielectric permittivity and magnetic permeability.

3.3.3 Let's Look at an Example under Research

In this section, a trial to develop low loss magneto-dielectric nanocomposite samples consisting of epoxy resin and iron nanoparticles is introduced in detail [4].

3.3.3.1 For what purpose?

Figure 3.15 shows the expected roles of constituent materials in composite of epoxy resin containing iron (Fe) nanoparticles. If this composite is used as a substrate of an antenna, it is expected that high μ and high dielectric strength can be attained by the presence of Fe nanofillers and epoxy resin, respectively.

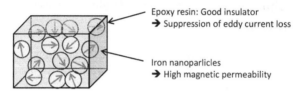

Figure 3.15 Concept of magneto-dielectric nanocomposite. Arrows stand for magnetic moments.

3.3.3.2 How to make the nanocomposite?

Diglycidyl ether of bisphenol A (816B, Mitsubishi Chemical) was used as a base epoxy resin and alicyclic polyamine (113, Mitsubishi Chemical) was used as a curing agent. Furthermore, alpha Fe particles with a catalog diameter of 70 nm were used as nanofillers to be added to epoxy resin.

For any nanocomposite, a complete uniform dispersion of unagglomerated nanofillers is most essential. Regarding this, ferromagnetic Fe particles tend to agglomerate. Therefore, oleylamine ($C_{18}H_{35}NH_2$) was used as a surfactant, since it has an amino group (–NH_2) with a tendency to be adsorbed easily onto the surfaces of nanoparticles and an alkyl group (–$C_{18}H_{35}$) which is easily dispersed in an organic solvent.

As a typical example, Fig. 3.16 shows the surface treatment procedures used to make this nanocomposite. First, the Fe particles, oleylamine as a surfactant, and zirconia beads for mixing were put in a container and the container was rotated in a planetary centrifugal mixer. By this procedure, agglomerations of Fe particles were broken up and the surface coating with oleylamine was completed. As a next step, the nm-sized Fe particles were added into uncured epoxy resin and the mixture was mixed in an ultrasonic homogenizer. Then, the liquid-like nanocomposite was injected into a mold and cured twice for adequate periods at proper temperatures.

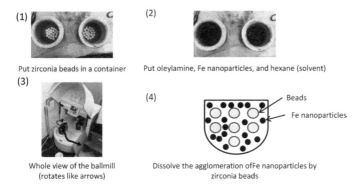

Figure 3.16 Procedure of surface treatment of Fe nanoparticles by a ballmill. [(1) → (3) and explanation (4)].

3.3.3.3 Let's look at important properties

(i) How good are the nanoparticles dispersed?

The dispersability of nanoparticles in the polymer is usually observed by transmission electron microscopy (TEM) or scanning electron microscopy (SEM). Figure 3.17 shows SEM images of the surfaces of the samples containing Fe nanoparticles. The samples containing the nanoparticles with contents of 20 and 40 vol% have large agglomerations, although agglomeration is not seen when the filler content is 6 vol%. Furthermore, Figs. 3.17d–f, which are magnified by 5000 times, indicate that the surfaces of nanoparticles seem to be coated with the resin.

(ii) Is the conductivity low enough?

Figure 3.18 shows the dependence of the electrical conductivity σ on the particle content p. At $p < 2.6$ vol% or in region I, σ increases hardly, while at $2.6 < p < 30$ vol% in region II, σ shows a very rapid increase with an increase in p. Then, in region III where $p > 30$ vol%, σ increases rather slowly. At the maximum content $p = 40$ vol%, σ is 1.2×10^{-5} S/m, which is lower than the value of ferrite ($\geq 10^{-4}$ S/m).

114 | Compatibility of Dielectric Properties with Other Engineering Performances

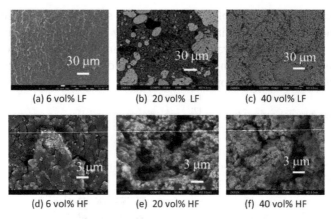

Figure 3.17 Cross-sectional SEM images observed for the samples with the particle contents of 6, 20 and 40 vol% [4]. LF: Taken with a low magnification of 500 times, HF: 5000 times.

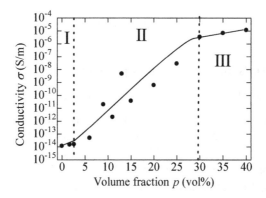

Figure 3.18 Conductivity σ as a function of particle content [4]. Roman numerals I, II, and III indicate three regions where σ shows different dependences. —: curve fitted to data.

(iii) How high is the permittivity?

Figure 3.19 shows frequency dependences of relative permittivity ε'_r and relative dielectric loss ε''_r, observed for the samples containing various contents of nanoparticles. The samples with p of 35 and 40 vol% show a very rapid increase both in ε'_r and ε''_r especially at low frequencies. It is clearly shown in Fig. 3.19 that both ε'_r and ε''_r at 10^5 Hz increase with an increase in p. The sample with the filler content of 40 vol% shows ε'_r of ~9.8, which is higher than ε''_r of epoxy resin.

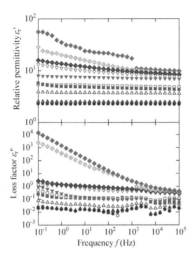

Figure 3.19 Relations between complex relative permittivity and frequency observed for each sample [4]. ●: 0 vol%, ○: 0.8 vol%, ▲: 1.7 vol%, △: 2.6 vol%, ■: 9.0 vol%, □: 13 vol%, ▼: 20 vol%, ∇: 25 vol%, ◆: 30 vol%, ◈: 35 vol%, ✦: 40 vol%.

As shown in Fig. 3.19, both ε'_r and ε''_r increase rapidly at low frequencies in the samples with p = 35 or 40 vol%. Since the permittivity of a conductor is infinite if the frequency is low, it is reasonable that the permittivity increases with the increase in the content of Fe nanoparticles. The permittivity ε'_r, 9.8, of the sample with p of 40 vol% at 10^5 Hz is fairly high for electrical insulating materials. Therefore, the attempt to raise ε'_r was successful. On the other hand, since conduction current generates Joule heat, high electrical conductivity increases ε''_r. Although ε''_r increases with the increase in p, the values of ε''_r at important high frequencies remains at 0.42 even at 40 vol% as shown in Fig. 3.19, which is an acceptable low level for electronic devices.

Furthermore, ε''_r does not increase so much in the samples with p < 30 vol%. This is seemingly because the increase in σ can be suppressed by coating the surfaces of Fe nanoparticles with oleylamine and by the good dispersion of the nanoparticles without showing heavy agglomeration.

(iv) What are magnetic properties?

Figure 3.20 shows the frequency dependences of the complex relative permeability μ'_r and μ''_r measured for the samples with nanofiller contents of 20 < p <40 vol%. With the exception that μ_r'' is higher at 35 vol% than at 40 vol%, both μ_r' and μ_r'' increase monotonically with the increase in p. Note that μ_r' was measured much lower than the correct values. While μ_r' decreases with the increase in frequency in a frequency range from 1.4 to 3.3 GHz, μ_r'' shows a peak at 2.25 GHz. This peak is due to ferromagnetic resonance. As is well known, a magnetic moment rotates due to precession in a magnetic field. This brings about a strong resonant absorption in a microwave frequency range, which is called ferromagnetic resonance and its frequency is called ferromagnetic resonance frequency (f_r). Since μ_r' decreases if the frequency is higher than f_r, f_r becomes the frequency limit usable for high frequency devices. While the present value of f_r = 2.25 GHz is fairly high, an even higher frequency is desirable for practical purposes.

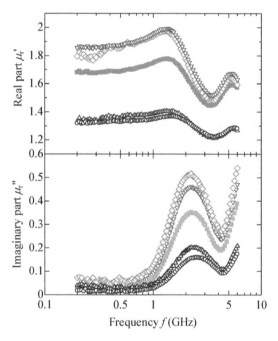

○: 20 vol%, Δ: 25 vol%, ■: 30 vol%, ◊: 35 vol%, ∇: 40 vol%.

Figure 3.20 Complex permeability as a function of frequency measured for five kinds of samples with relatively high particle contents [4].

Furthermore, the fact that no other peak than the ferromagnetic resonance peak is seen indicates that eddy current is not induced. Namely, the surface coating is effective to prevent the eddy current. The eddy current generally induces a new magnetic field that would prevent the change in the magnetic field, which decreases μ_r'. Therefore, the suppression of eddy current is important for the development of a low loss magneto-dielectric materials.

References

1. An, Y., Nishida, K., Yamamoto, T., Ueda, S., and Deguchi, T. (2008). Microwave Absorber Properties of Magnetic and Dielectric Composite Materials, *IEEJ Trans. Fundamentals Mater.*, **128**(6), pp. 441–448 (in Japanese).

2. Hasegawa, D., Ogawa, T., and Takahashi, M. (2009). Nanoparticle-Based Magneto-Dielectric Hybrid Material for High-Frequency Devices, *Magn. Japan*, **4**(4) pp. 180–185 (in Japanese).

3. Borah, K., and Bhattacharyya, N. S. (2010). Magneto-dielectric Material with Nano Ferrite Inclusion for Microstrip Antennas: Dielectric Characterization, *IEEE Trans. Dielectr. Electr. Insul.*, **17**(6), pp. 1676–1681.

4. Hirose, Y., Hasegawa, D., and Ohki, Y. (2015). Development of Low Loss Magnetodielectric Nanocomposites of Epoxy Resin and Iron Nanoparticles, *Electrical Engineering in Japan*, **190**(2) (Translated from *IEEJ Trans. Fundamentals and Materials*, **133**, No. 12, pp. 668–673, 2013).

5. Ariake, J., Chiba, T., and Honda, N. (2005). Co-Pt-TiO_2 Composite Film for Perpendicular Magnetic Recording Medium, *IEEE Trans. Magn.*, **41**(10), pp. 3142–3144.

6. Sellmyer, D. J., Luo, C. P., Yan, M. L., and Liu, Y. (2001). High-Anisotropy Nanocomposite Films for Magnetic Recording, *IEEE Trans. Magn.*, **37**(4), pp. 1286–1291.

7. Zhang, Y. D., Wang, S. H., Xiao, D. T., Budnick, J. I., and Hines, W. A. (2001). Nanocomposite Co/SiO_2 Soft Magnetic Materials, *IEEE Trans. Magn.*, **37**(4), pp. 2275–2277.

8. Yang, T.-I., Brown, R. N. C., Kempel, L. C., and Kofinas, P. (2008). Magneto-dielectric Properties of Polymer–Fe_3O_4 Nano-composites, *J. Magn. Magn. Mater.*, **320**(21), pp. 2714–2720.

3.4 High-Heat-Resistant Composites

Norio Kurokawa

Japan Electrical Insulating and Advanced Performance Materials Industrial Association, Sumida-ku, Tokyo 130-0014, Japan

jeia@vesta.ocn.ne.jp

The heat resistance of polymers is expected to improve by nanocomposite technology. Raising of the glass transition temperature (T_g) and improvement in the pyrolytic property by nanocomposites have been reported. This section describes the improvement in the heat resistance of polymers and dependence on the dispersion method.

3.4.1 Research Is Progressing in High-Heat-Resistant Composites by Using Nanocomposites

Polymer nanocomposites have been studied by many researchers, and an increasing of the strength modulus at high temperature and higher T_g has been reported [1]. Heat resistance improvement of polymers by using nanocomposites is expected in industrial applications. Therefore, research is still being carried out on the combination of various polymers and nanocomposites.

3.4.1.1 Polyethylene

The effect of added SiO_2 to cross-linked polyethylene (XLPE) has been studied [2]. The heat degradation was indicated by generation of a carbonyl group, and the amount of the carbonyl group was observed by the ATR method of the Fourier transfer infrared spectrometer (FT-IR). The examination was made by comparison between the case of adding filler only and the case of antioxidant–filler combination. The change in the generated quantity of the carbonyl without antioxidant at 130°C is shown in Fig. 3.21. The change in the generated quantity of carbonyl by adding 0.2% antioxidant at 180°C is shown in Fig. 3.22. It used carbonyl absorbance 0.05 as an indicator of degradation. Lifetime of polyethylene without antioxidant is 50 h at 130°C, but the lifetime can be extended to 80 h at 130°C by adding 0.2 wt% nano-sized SiO_2 filler.

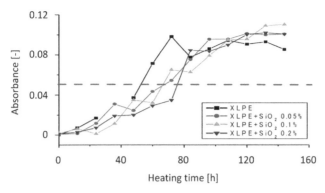

Figure 3.21 Absorbance change in SiO$_2$ addition XLPE at 130°C [2].

Further, by adding 0.2 wt% antioxidant and 0.1 wt% of nano-sized filler, the thermal degradation at 180°C is greatly extended to 100 h from 20 h. Thus, the effect of the addition of the nano-sized filler is observed for thermal degradation also. Also, by combining an antioxidant, heat degradation of polypropylene extends more significantly. In this way, improvement in heat resistance is observed due to the combination of a nanofiller and another ingredient. That is an important means in the industrial sector.

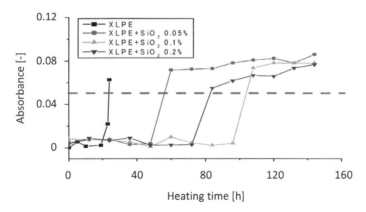

Figure 3.22 Absorbance change in SiO$_2$ addition XLPE at 180°C [2].

3.4.1.2 Silicone resin

Heat resistance improvement by the adding the nanofiller in the silicone resin has been studied [3]. The weight loss of silicone resin with adding nano fumed silica is smaller than silicone resin adding micro silica. In case of silicone resin which added micro silica, the weight loss becomes 70% at 600°C.

On the other hand, in case of silicone resin with added nano fumed silica, the heating weight loss becomes 30% at 600°C. In this way, the further improvement of heat resistance is realized by making nanocomposite silicone resin as well.

3.4.1.3 Epoxy resin

It is studied the effect of adding nanofiller composite to epoxy resin. By the addition of nano-silica, the glass transition temperature is decreased approximately 10°C [4]. The change in the glass transition temperature is different depending on the mixing ratio of the hardener.

Regarding the nanocomposite of the epoxy resin, many reports have shown improvement in heat resistance [1]. However, dependence on the dispersion state of the resin-hardener-nanofiller decreased heat resistance. In each application field, it is necessary to consider the polymer nanomaterials and dispersion method in order to develop high-heat-resistant nanocomposites.

3.4.2 Thermal Endurance Varies with Methods of Nanofiller Dispersion

The glass transition temperature of epoxy/alumina nanocomposites has been studied [5]. Even in the same epoxy and amine hardened, the T_g was different depending on the dispersion method. Those of epoxy resin dispersed in a distributed well using high-pressure homogenizer, T_g compared to the dispersion with normal revolving mixer became 4°C higher. Thus, method to perform the dispersion of the resin and nanofillers is an important factor. Silicone rubber, nano silica, nanocomposite had been created with mechanical and electro spinning method by S. Bian et al. [6]. Electrospinning method is a method of increasing the spinning by dispersing the polymer from the narrow tip of the needle multiplied by the electric field. The silicone rubber which contained 20% micro silica and 5% nano silica was made by the normal machine stirring and electrospinning method.

Therefore, silicone rubber's characteristics comparison was done based on a different method.

The thermal decomposition characteristics were measured with thermogravimetric analysis (TGA). Whereas the weight loss at 700°C in case of the normal mechanical stirring was 35%, the loss became 31% in case of dispersion by the electrospinning method. The difference of 4% thermal decomposition was observed by the dispersion method. Thus, the thermal decomposition differences were confirmed by the dispersion method, even after the addition of the same nano silica.

Thus, many studies have been carried out on the nano-dispersion method also. It has been reported that there is some difference in the heat resistance of the nanocomposite by the difference of the dispersion method.

It is important whether the dispersion of nanofiller components is done at the nano level.

3.4.3 Practical Implementation of High-Thermal-Endurance Composites Is Now in Progress

The development of the nanocomposite of epoxy resin and silica has been carried out by using sol-gel techniques using alkoxysilane [7]. Figure 3.23 shows the basic structure of alkoxysilane-bonded hybrid epoxy resin. In addition, Fig. 3.24 shows the structure diagram of the silica hybrid resin after sol-gel curing the alkoxysilane-bonded hybrid epoxy resin. The alkoxysilane groups were introduced to the thermally weak site of the bisphenol-based epoxy resin in the structure and then a high-heat resistance hybrid material of epoxy resin and nano-sized silica was formed after sol-gel curing.

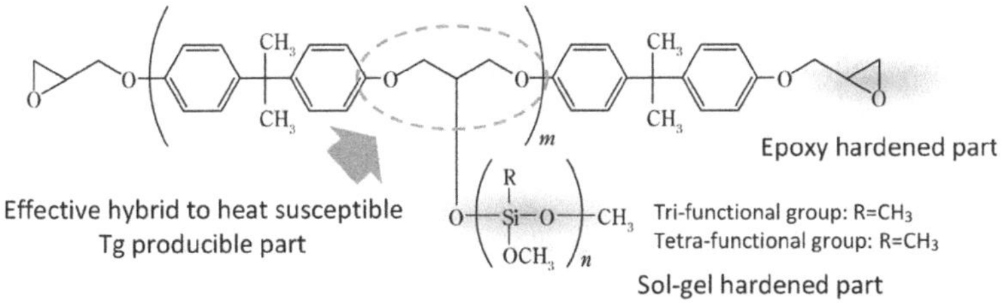

Effective hybrid to heat susceptible
Tg producible part

Epoxy hardened part

Tri-functional group: R=CH₃
Tetra-functional group: R=CH₃

Sol-gel hardened part

Figure 3.23 Alkoxysilane-bonded hybrid epoxy resin [7].

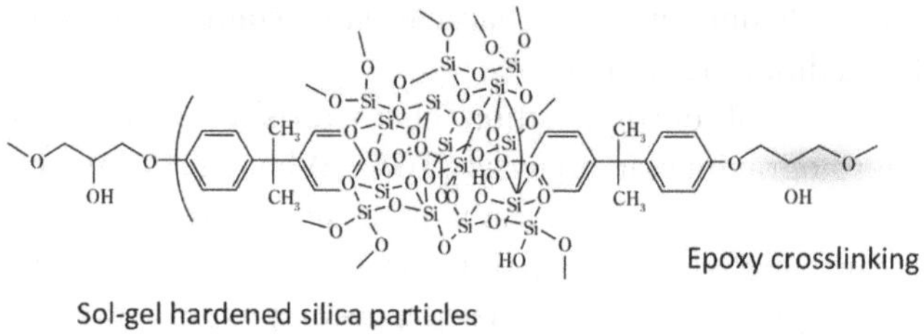

Figure 3.24 Imaging of epoxy resin-silica hybrid after curing [7].

The dynamic viscoelasticity of the cured product of the hybrid material is shown in Fig. 3.25. The figure shows T_g has disappeared. In addition, even at the high temperature over 150°C or more, the storage modulus coefficient of elasticity did not decrease. Thus, a nanocomposite can achieve a significant improvement in the storage modulus at high temperature.

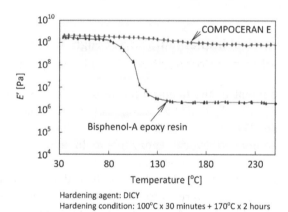

Figure 3.25 Dynamic viscoelasticity of COMPOCERAN E [7].

The techniques of this nanocomposite have been applied to phenolic resins, polyimide resins, polyamide-imide resins, acrylic resins, siloxane resins, and polyimide film. These nanocomposite materials have established improvement in various characteristics such as heat resistance, improvement in adhesion, reduction of C.T.E., etc.

EVA (ethylene-vinyl acetate copolymer) is superior in transparency, flexibility, and adhesiveness. Therefore, EVA is widely used for solar cells, films, sheets, laminates, and in fields such as adhesives. Nanocomposites of EVA have been studied.

A high softening temperature EVA nanocomposite was developed [8]. The structure of the EVA nanocomposite is shown in Fig. 3.26. The dynamic viscoelasticity of data is shown in Fig. 3.27. Storage modulus and softing point of EVA rise by becoming the nanocomposite.

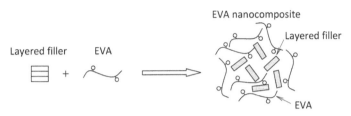

Figure 3.26 The structure of EVA nanocomposite [8].

Nylon 6 with nanocomposite has been developed [9]. The nanocomposite was formed using a layered silicate. Layered silicate was split into more by one part, and it was subjected to polymerization by dispersing the layered silicate into nylon. In this way, the polymer with excellent heat resistance could be obtained, compared with nylon 6 polymer. Thus, the strength, storage modulus, bending strength, and bending modulus of nylon nanocomposite could be improved compared with normal nylon at the high temperature. The tensile strength of normal nylon is reduced 25% at 100°C and 1000 h. In comparison, in nanocomposite nylon, the tensile strength is hardly decreased.

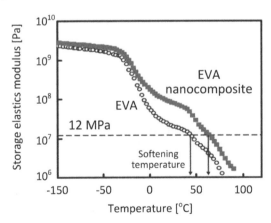

Figure 3.27 Temperature dependence of the storage elastic modulus (acetic acid vinyl content 28%) [8].

Heat resistance of the epoxy resin has been improved by adding a few percent of layered silicate. Figure 3.28 shows a TEM photograph and a schematic view of nanocomposite epoxy resin. It is observed that resin impregnates between the layers of the layered silicate. Figure 3.29 shows a result of dynamic mechanical analysis (DMA). The nanocomposite of epoxy resin exhibits higher storage elastic modulus and higher Tg than the original epoxy resin. Thus, the resin's characteristics could be improved by nanocomposite technology. The technological advance of the insulating material for the electricity apparatus has been advanced by the application.

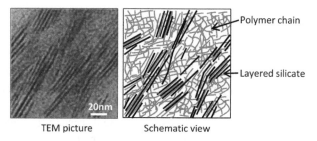

TEM picture   Schematic view

Figure 3.28   Prepared nanocomposite [10].

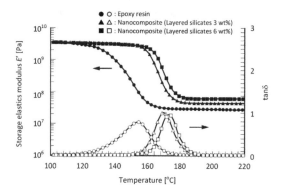

Figure 3.29   Thermal dependences of E' and tan δ from dynamic mechanical analysis (DMA) measurements [10].

According to the advances in the development of SiC devices in recent years, the heat resistance requirements are becoming higher. SiC power devices have superior performance compared with silicon power devices. Therefore, high-efficiency small power devices can be developed while applying the SiC technology.

Silicone devices are limited to operate at temperature under 150°C. In contrast, SiC devices can operate at even 400°C or higher. Therefore, modules that use SiC devices can be operated at higher than 200°C.

Therefore, in order to protect the SiC elements, a resin with higher heat resistance is required. Nanocomposites also have been studied as a method of increasing heat resistance of the resin. NIPPON SHOKUBAI developed a nanomaterial of silane with hydrolysis-condensation reaction of organometallic compounds. In addition, they granted a high-affinity functional group to nanomaterials in the thermosetting resin. The thermosetting resin has a chemical bond with nanomaterials. By such means, they developed a nano-hybrid resin, which was made with a thermosetting resin as a matrix [11, 12].

The nanocomposite of the epoxy resin with this structure achieved significant improvement in heat resistance. In addition, nano-hybrid methods involve molecules designed to be applicable to various thermosetting resins such as epoxy resin or phenolic resin. By using this technique, low melting point and liquefaction even with highly crystalline epoxy resin were observed. Therefore, this technique can be applied for the development of liquid molding compounds. Moreover, the possibility of realizing an IPN (inter penetration network) structure of nano level has also been suggested. Figure 3.30 shows the Si distribution of the nanocomposite. It is observed that Si is uniformly dispersed at the nano level. Figure 3.31 shows the morphology of the nano level in the nano-hybrid resin. The grape-like substance shows nanomaterials, and the ribbon-like material shows the epoxy resin.

Figure 3.30 Si element mapping of nanocomposite [11].

Figure 3.31 Concept image of nanocomposite [11].

The interaction of nanomaterials and the epoxy resin forms a very close network structure. For these structures, thermal decomposition characteristics are greatly improved. Figure 3.32 shows the weight loss data on heating in air at 200°C, which is the epoxy resin used as the matrix of nano-hybrid resin cured product.

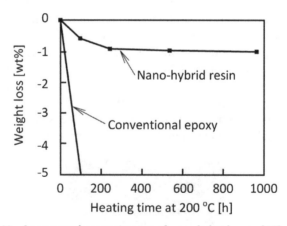

Figure 3.32 Long-term heat resistance of nano-hybrid resin [11].

By this technique, it has been observed that the thermal decomposition resistance is improved and long-term heat resistance stability at temperatures higher than 200°C is also observed. Using this approach, Nippon Shokubai develops higher-heat-resistance molding compounds and printed circuit board material, etc. [12].

There are several examples of heat-resistant polymer development by using the nanocomposite technique. By using nanocomposites, practical research to improve heat resistance of the polymer has been developed by many companies. Heat resistance has been improved similar to increasing the glass transition temperature, softening point of the resin material, and improving the thermal elastic modulus by applying nanocomposite.

The nanocomposite is expected to be applied to the fields of molding compounds, electrical materials, and electronic materials. These improved characteristics depend on the nanofiller dispersion method. Therefore, the dispersion method is an important development theme for nanocomposites. Nanocomposites help not only improve physical properties but also enhance the thermal decomposition temperature. In case of epoxy resins, the development of nanocomposites with higher heat-resistant properties at 200°C or higher temperatures is expected.

References

1. Ochi. M. (2004). Epoxy Resin Nanocomposite, *J. Adhes. Soc. Japan*, **40**(4), pp. 37–41 (in Japanese).

2. Gondo, Y., Noguchi, K., and Maeno, T. (2010). Research on Thermal and Electrical Specification of SiO_2 Nanofiller Added XLPE, *Proc. IEEJ Natl. Convention*, **2**(2–033), pp. 38–39 (in Japanese).

3. Ramirez, I., Cherney, E. A., Jayaram, S., and Gauthier, M. (2008). Thermogravimetric and Spectroscopy Analyses of Silicone Nanocomposites, *Proceedings of the IEEE CEIDP*, pp. 249–252.

4. Nguyen, A., Vaughan, A. S., Lewin, P. L., and Krivda. A. (2011). Stoichiometry and Effects of Nano-sized and Micro-Sized Fillers on an Epoxy Based System, *Proc. IEEE CEIDP*, **1**(3B-1), pp. 302–305.

5. Hase, Y., Kosako, M., Ootuka, S., Hikita, M., and Tanaka, T. (2009). Making of Epoxy/Almina Nanocomposite and Observation of T_g, *Proc. IEEJ Natl. Convention*, **2**(2–028), p. 35 (in Japanese).

6. Bian, S., Jayaram, S., and Cherney, E. A. (2012). Electrospinning as a New Method of Preparing Nanofilled Silicone Rubber Composites, *IEEE Trans. Dielectr. Electr. Insul.*, **19**(3), pp. 777–785.

7. Arakawa Chemical Industries, LTD. (2012). *COMPORACEN Technical Data*, pp. 1–13 (in Japanese).

8. Senba, M., Mori, K., and Yukioka, S. (2010). Characteristics of EVA Nanocomposite, *TOSOH Res. Technol. Rev.*, **54**, pp. 41–46 (in Japanese).

9. UNITIKA LTD. (2012). *Nanocomposite Nylon6, Technical Data.*

10. Ozaki, T., Imai, T., and Shimizu, T. (2004). Functional Insulating Materials Using Nanoparticle Dispersion Technique, *TOSHIBA Rev.*, **59**(7), pp. 48–51.

11. Nippon Shokubai Co., LTD. (2013). *High Heat Resistant Nano-hybrid Resin, Technical Data*, p. 1.

12. Nippon Shokubai Co., LTD. (2012). *Molding Compound for SiC Power Device, Technical Data*, pp. 1–2.

3.5 Composites with High or Low Permittivity

Kenji Okamoto

Fuji Electric Co., LTD., Hino-shi, Tokyo 191-8502, Japan

okamoto-kenji@fujielectric.co.jp

In the electronics field, there is an increasing need for downsizing and high-density mounting. That is why high-κ (high-dielectric constant) and low-κ (low-dielectric constant) materials need to be developed urgently as LSI circuit elements and LSI materials. Today's material developments focus on high-dielectric inorganic materials like Hf (hafnium)-based oxides as high-κ materials, and porous ceramics as low-κ materials, but polymer composites are drawing attention from the viewpoints of both processability and flexibility [1].

3.5.1 Are Composites with High or Low Permittivity?

This section describes why high-dielectric-constant materials (high-κ) and low-dielectric-constant materials (low-κ) are needed for the downsizing and high-density mounting of LSI (large-scale integrated) circuits, and how they will be used for actual applications.

3.5.1.1 High-dielectric-constant materials

High-dielectric-constant materials are used as insulator films for semiconductor gate electrodes. An oxide film isolates the gate electrode from the semiconductor, and electrical signals are transmitted by electrostatic induction. If two oxide films have the same thickness, the one with a higher dielectric constant can transmit an electrical signal at a faster speed. Because leakage current limits how thin oxide films can possibly be, high-κ materials need to be developed.

Until today, the improvements in the performance of Si (silicon)-based LSI have contributed significantly to rapidly advancing information processing technology. The high performance has been achieved through the miniaturization of metal oxide semiconductor (MOS) transistors (Fig. 3.33), which is the basic structural unit of LSI. As MOS transistors become more miniaturized, the thickness of their gate insulator film should be thinned incrementally. Thanks to miniaturization, capacitance per unit area C_1 becomes higher, and the drive current of a transistor increases. When the dielectric constant and the film thickness of a gate insulator film are ε and d, respectively, C_1 is given by

$$C_1 = \frac{\varepsilon}{d} \tag{3.5}$$

Note that dielectric constant f is the product of the relative permittivity and the vacuum permittivity. Relative permittivity is expressed by ε or κ, and particularly in the LSI field, κ seems to be used more commonly.

When the film thickness is 1 nm or thinner, conventional gate insulator films made of silicon oxide (SiO_2) lose their insulating properties, because leakage current is induced by the quantum tunnel effect. This phenomenon leads to LSI's power consumption well in excess of the allowable limit. One of the solutions to this problem is the use of high-dielectric-constant (hereinafter referred to as "high-κ") gate insulator film technology. The use of insulator film materials with a larger f value than SiO_2 could make the C_1 value larger without decreasing the d value, and hence increasing drive current while reducing the amount of leakage current.

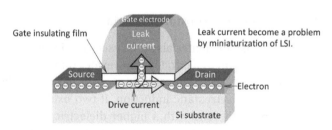

Figure 3.33 Structure of a MOS transistor [2].

According to the International Technology Roadmap for Semiconductors (ITRS), which predicts progress in semiconductor product technology by generation, Hf-based oxides, such as hafnium silicate (HfSiO) and nitrogen-added hafnium silicate (HfSiON), will likely be applied to LSI chips around 2008 when high-κ gate insulator film is mounted for the first time. Hf-based oxides have high heat resistance and are compatible with conventional LSI processes. Additionally, this generation requires an equivalent oxide thickness (EOT) of a little less than 1 nm, which Hf-based oxides can satisfy.

On the other hand, it is also predicted that the gate insulator films for semiconductors whose mass production is scheduled for sometime around 2016 must have an EOT of 0.5 nm or thinner. This is why materials with a higher dielectric constant will be needed (Fig. 3.34) [2].

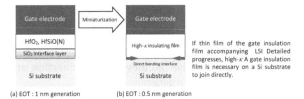

Figure 3.34 Change of the structure of a High-κ gate insulation film by miniaturization of LSI [2].

3.5.1.2 Low-dielectric-constant materials

Low-dielectric-constant materials are used as interlayer dielectrics in LSI chips. This is because materials with a lower dielectric constant are needed to suppress electrostatic induction and increase the independency of each layer. Those materials could reduce signal propagation delay, which poses a big problem in the process of miniaturization. Since the 130 nm generation of high-speed logic devices, a new material has been adopted until now in the multilayer wiring process for LSI chips of each generation. As for the wiring material, Cu (copper) wire has begun to be used in place of Al (aluminum) wire, which had been used since the 130 nm generation until the previous generation. At the same time, since the 90 nm generation, low-dielectric-constant films (Low-κ films) with a relative permittivity of 3.0 or lower have been used as interlayer dielectric film. Since then, the dielectric constants of interlayer dielectric films have continued to decline. In the 65 nm generation and later, those films should have micro voids inside and have an electric constant of 2.5 or lower. Figure 3.35 shows its schematic view [3].

As discussed above, the downsizing and high-density mounting of LSI circuits would require materials with higher/lower dielectric constants. It is therefore necessary to not only refine existing materials but also pursue innovative material development from a new perspective. As a novel idea, many expect the emergence of nanocomposites, combinations of a various types of resins and nanofillers (nanoparticles).

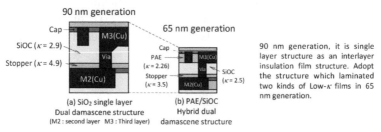

Figure 3.35 The outline of the multilevel interconnection structure in 90 nm and a 65 nm generation [3].

3.5.2 Permittivity Can Be Increased by Dispersion of High-Dielectric-Constant Nanofiller

A common method for increasing the dielectric constant of a resin is to disperse a selected high-dielectric-constant filler into the resin. The following are specific cases where the addition of a nanofiller (nanoparticle) increased the dielectric constants of resins.

Here is how to prepare nanosized barium titanate ($BaTiO_3$). In the first step as shown in Fig. 3.36, after being heated in the atmosphere at 500°C, $BaTiO(C_2O_4)_2$ will be decomposed into $BaCO_3$ and TiO_2. The first step is followed by the second step in which reaction products are heated in vacuum at a high temperature (630 to 830°C), producing $BaTiO_3$ nanoparticles. The particle size can be controlled by a change in the temperature condition (in the range of 630 to 830°C).

Figure 3.37 shows TEM images and grain sizes of $BaTiO_3$ nanoparticles prepared under different conditions. Depending on the heating temperature and condition in the second step, the nanoparticles can be made in the diameter range of 17 to 68 nm. Note that A-4 and A-5 in the figure were heated at the same second step temperature (830°C) but in different atmospheres: in vacuum and in the atmosphere. And the resulting nanoparticle diameters were different, 68 and 102 nm, respectively.

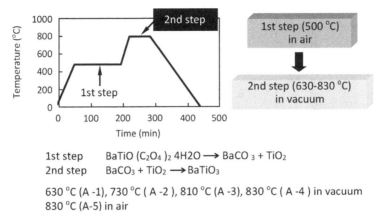

1st step   BaTiO (C$_2$O$_4$)$_2$ 4H2O ⟶ BaCO$_3$ + TiO$_2$
2nd step   BaCO$_3$ + TiO$_2$ ⟶ BaTiO$_3$

630 °C (A -1), 730 °C (A -2), 810 °C (A -3), 830 °C (A -4) in vacuum
830 °C (A-5) in air

Figure 3.36 The particle production conditions of barium titanate (BaTiO$_3$) [4].

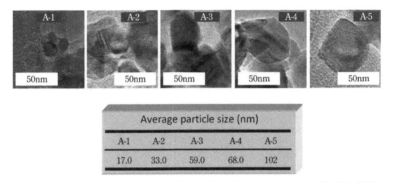

Figure 3.37 Particle diameter of the produced barium titanate (BaTiO$_3$) [4].

The relative permittivity of these barium titanates with a different particle diameter were analyzed by the finite element method (FEM). As a result, the relative permittivity increased with the decreasing particle diameters, reaching a maximum of 15,000 when the grain size was about 68 nm (Fig. 3.38). High-dielectric-constant, 68 nm barium titanate nanoparticles were modified on the surface by a material called hyperbranched phthalocyanine, which has a dendritic structure as shown in Fig. 3.39. Then these nanoparticles were dispersed into a polyamide (PA) in concentration as high as 70 vol%. Despite being a polymer composite, this material achieved a relative permittivity of 80 at 1 MHz (Fig. 3.40) [4].

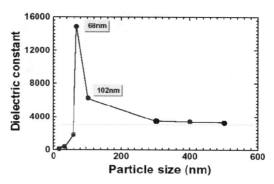

Figure 3.38 Relation between the diameter of a particle by FEM, and a dielectric constant [4].

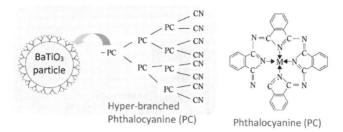

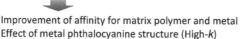

- Phthalocyanine complex formation with Ba^{2+}, Ti^{2+} of $BaTiO_3$
- Surface covered by cyano groups

Improvement of affinity for matrix polymer and metal
Effect of metal phthalocyanine structure (High-k)

Figure 3.39 The surface modification by hyper-brunch phthalocyanine [4].

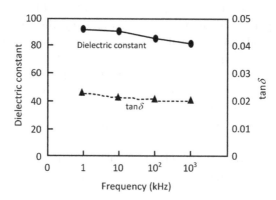

Figure 3.40 The relative permittivity of nanocomposite [4].

This accomplishment is attributable to the finding that the relative permittivity of barium titanate, which intrinsically has a high permittivity, can be increased further by making its particles smaller down to nanosize, as well as the successful development of a nanocomposite by modifying the nanoparticle surface so that they can be dispersed into a polymer in high concentration.

3.5.3 Permittivity Can Be Lowered by Low-Permittivity Nanofiller

Generally speaking, materials used as fillers have a high permittivity. If such filler is dispersed into a polymer, the polymer's permittivity becomes higher. For this reason, there are efforts to try to reduce dielectric constants using hollow fillers, whose particles have voids inside (the dielectric constant of air is as low as 1).

The dielectric constant of the nanocomposite made of 4 nm-diameter mesoporous silica (MCM-41) dispersed into a polyimide (PI) depends on the amount of the nanoparticle. The permittivity becomes the lowest of 2.58 (at 103 Hz and 25°C) when 3 wt% of MCM-41 is added, a considerable reduction from the polyimide's relative permittivity of 2.94 (Fig. 3.41) [5]. The calculated relative permittivity (theoretical value) of the nanocomposite, which took the voids in the nanoparticle into consideration, is higher than its experimental value. This suggests that a number of factors other than the hollow particle structure may contribute to the reduction in permittivity.

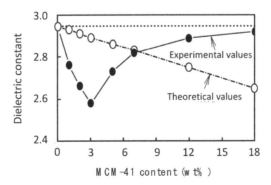

Figure 3.41 The example of low permittivity-ized by nanocomposite (1) [5].

Meanwhile, in comparison with a composite made of alumina microparticles (10 μm in particle diameter) dispersed into an epoxy resin, a nanocomposite made of the same weight percent of nanoparticles dispersed into the same epoxy resin has a lower permittivity. As the amount of added particles increases, the difference in permittivity between the two composites tends to decrease (Fig. 3.42) [6].

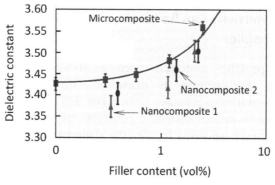

<Dispersion method of fillers>
Nanocomposite 1: High-speed planetary mixer
Nanocomposite 2: High-speed planetary mixer + ultrasonic wave

Figure 3.42 The example of low permittivity-ized by nanocomposite (2) [6].

The reduction in permittivity dielectric constant by the use of a nanocomposite is considered attributable to the influence of the area restricted by molecular chains (low-permittivity dielectric constant layer) due to strong particle-base material bonds on the nanoparticle interface. Using an FEM-based current analysis, capacitive current flowing through the model as shown in Fig. 3.43 was calculated. In this model, alumina nanoparticles are dispersed into an epoxy resin and a low-permittivity C layer is formed on the nanoparticle interface. The capacitance of the capacitor (C) was obtained from the formula shown in the figure, the permittivity of the nanocomposite was calculated, and then the lower theoretical limit of dielectric constant was studied.

As result, it is possible that the dielectric constant of the nanocomposite is 13% lower than that of the neat epoxy resin, when the low-permittivity layer's ratio of thickness to particle radius is 1.5, and when ε NC/ε epoxy becomes 0.87 with the nanoparticle dispersed to 16 wt%, as shown in Fig. 3.44 [7].

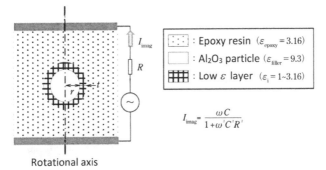

Figure 3.43 Current analysis model [7].

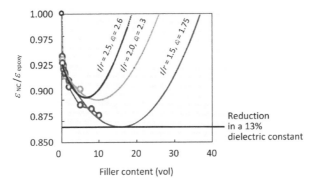

Figure 3.44 Filler content dependence property of low permittivity-ized effect [7].

References

1. Tanaka, T. (2009). Suggested Study on Advanced Composite Materials: Super Composites—Development of Technology for Dielectric Isolation Materials, *Proc. the 40th IEEJ ISEIM*, No. B-1, pp. 33–38.

2. Suzuki, M., Yamaguchi, T., and Koyama, M. (2007). Lanthanum Aluminate Gate Dielectric Technology with Direct Interface, *Toshiba Rev.*, **62**(2), pp. 37–41 (in Japanese).

3. Yoda, T., Hasunuma, M., and Miyajima, H. (2004). Advanced BEOL Technology, *Toshiba Rev.*, **59**(8), pp. 17–21 (in Japanese).

4. Takahashi, A., Kakimoto, M., Tsurumi, T., Hao, J., Li, L., Kikuchi, R., Miwa, T., Ohno, T., Yamada, S., and Takezawa, Y. (2006). Polymer-Ceramic Nanocomposites Based on New Concepts for Embedded Capacitor, *IEEJ Trans. Fundamentals Mater.*, **126**(11), pp. 1160–1166.

5. Dang, Z., Song, H., Lin Y., and Ma, L. (2008). High and Low Dielectric Permittivity Polymer-Based Nanohybrid Dielectric Films, *Proc. IEEJ ISEIM*, No. P1-33, pp. 315–318.

6. Kurimoto, M., Kai, A., Watanabe, H., Kato, K., and Okubo, H. (2008). Evaluation of Relative Permittivity of Epoxy/Alumina Nanocomposites Based on Grain Boundary Area, *Proc. IEEJ ISEIM*, **2**(2–052), p. 61.

7. Nakano, T., Hayakawa, N., Hanai, M., Kato, K., Sato, M., Hoshina, Y., Takei, M., and Okubo, H. (2012). Thoughts about Lower Limit of Relative Permittivity of Epoxy/Alumina Nanocomposites, *Proc. IEEJ ISEIM*, **2**(2–077), p. 93.

PART 2
FUNDAMENTALS
(MATERIAL PREPARATION)

Chapter 4

Preparation of Polymer Nanocomposites: Key for Homogeneous Dispersion

Mikimasa Iwata[a], Yuki Honda[b], Takahiro Imai[c] and Minoru Okashita[d]

[a]*Central Research Institute of Electric Power Industry (CRIEPI)*
[b]*Hitachi Metals, Ltd.*
[c]*Toshiba Corporation*
[d]*SWCC Showa Cable Systems Co., LTD.*

4.1 Reactive Precipitation Method: Sol-Gel Method

Mikimasa Iwata

Central Research Institute of Electric Power Industry (CRIEPI), Yokosuka-shi, Kanagawa 240-0196, Japan

m-iwata@criepi.denken.or.jp

A sol is a colloid in which the dispersion medium is a liquid and the dispersed material is a solid. A gel is a jelly-like material obtained by solidifying a sol. In the sol-gel method, a solution is converted into a gel via a sol, and glassy fine particles are

---

Advanced Nanodielectrics: Fundamentals and Applications
Edited by Toshikatsu Tanaka and Takahiro Imai
Copyright © 2017 Pan Stanford Publishing Pte. Ltd.
ISBN 978-981-4745-02-4 (Hardcover), 978-1-315-23074-0 (eBook)
www.panstanford.com

synthesized and dispersed in polymers. The sol-gel method has an advantage that nanofillers (nanoparticles) are dispersed more homogeneously than in the filler mixing method. Polymer nanocomposites are prepared using the sol-gel method and are applied practically as adhesives, coating agents, and electronic materials.

4.1.1 The Sol-Gel Method Can Result in Excellent Dispersion of Nanofillers in Polymers [1, 2]

In general, in the sol-gel method a glass is synthesized by heating a gel converted from a solution via a sol. Because a sol is a colloid, in which the dispersion medium is a liquid and the dispersed material is a solid, a sol is also called a colloidal solution. A sol is prepared by a method in which particles are grown due to condensation, deposition, and a reaction, or by a method in which coarse particles are made finer by mechanical, electrical, or chemical process.

The gel is a jelly-like material obtained by solidifying the sol. Although the gel contains vacant spaces or liquid components similarly to much water, the shape of the gel is well maintained because the gel has a support structure throughout its system. Examples of gels are agar, gelatin, soybean curd, and silica gel. Most sols have structural viscosity because of the attraction between colloidal particles dispersed in the sol. As the density of a sol increases more and more, the particles become connected with each other, three-dimensional (3-D) netlike or beehive-like structure is formed, and finally the dispersed system is converted into a solid-like gel.

By using the sol-gel method, nanofillers (nanoparticles) can be synthesized and dispersed in polymers, and polymer nanocomposites can be prepared. The basic reaction equations are shown in Fig. 4.1. Specifically, a metal alkoxide, such as an alkoxysilane, is hydrolyzed and condensed using chlorine as a catalyst. In general, tetraethoxysilane (TEOS) or tetramethoxysilane (TMOS) is used as the metal alkoxide and silica (SiO_2) nanoparticles are synthesized in the polymers. Although the sol-gel method has a disadvantage that large quantities of water and alcohol are necessary, the method has an advantage that nanoparticles are dispersed more homogeneously than in the filler mixing method.

Hydrolysis (using acid catalyst or chlorine catalyst)

$$Si(OR)_4 + H_2O \longrightarrow (OR)_3SiOH + ROH$$

$$\equiv SOR + H_2O \longrightarrow \equiv SiOH + ROH$$

Condensation (using acid catalyst or chlorine catalyst)

$$\equiv SiOH + HOSi\equiv \longrightarrow \equiv SiOSi\equiv + H_2O$$

Figure 4.1 Basic reaction equations of sol-gel method [3].

4.1.2 Good Recipe and Special Care Are Needed for the Sol-Gel Method

There are various methods for preparing polymer nanocomposites. First, this section introduces a method for preparing polymer nanocomposites made from polydimethylsiloxane (PDMS) with both good mechanical properties and high heat-resistivity [4]. Liquid silanol-terminated PDMS with an average molecular weight of 20,000 is used as the precursor for the organic component, and TEOS [$Si(OC_2H_5)_4$] is used as the precursor for the inorganic component.

Figure 4.2 shows one method for preparing the sol. Liquid A is obtained by mixing PDMS, TEOS, and 2-ethoxyethanol ($C_4H_{10}O_2$) in the mole ratio shown in Table 4.1. Liquid B is obtained by mixing acetic acid (CH_3COOH), water (H_2O), and $C_4H_{10}O_2$ in the mole ratio shown in Table 4.1. Liquid A is heated to 90°C by mixing with a hot stirrer. After the temperature of liquid A reaches 90°C, liquid B is added dropwise to liquid A. The solution is then mixed for 30 min at 90°C. Then, the mixed solution is allowed to cool to room temperature.

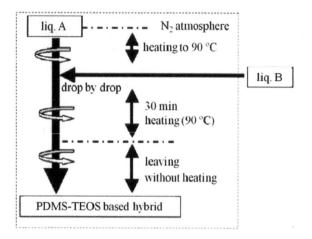

Figure 4.2 Experimental procedure for preparing TEOS-PDMS based hybrid sol [4].

The above preparation is carried out in a glove box, where the relative humidity is maintained at about 20% by replacing the atmosphere with dry nitrogen gas. The solution is poured into a Teflon Petri dish and covered with aluminum foil. The solution is cured by heating at 150 and 250°C for the times shown in Fig. 4.3. In this method, the gelation conditions around 100°C is one of the factors influencing the mechanical properties of cured polymer nanocomposites [5].

Table 4.1 Molecular weight and mole ratio of used materials for preparing the hybrid sol [4]

Liquid	Material	Chemical formula	Molecular weight	Mole ratio
A	PDMS	—	20,000	0.1, 0.2, 0.5, 1
	TEOS	$Si(OC_2H_5)_4$	208.33	1
	2-ethoxy ethanol	$C_4H_{10}O_2$	90.1	2
B	Water	H_2O	18	4
	acetic acid	CH_3COOH	60.65	0.05
	2-ethoxy ethanol	$C_4H_{10}O_2$	90.1	1

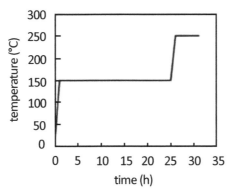

Figure 4.3 Curing temperature of the TEOS-PDMS nanocomposites [4].

Next, a recent study on the preparation of polymer nanocomposites made from polypropylene (PP) is introduced. There is a problem that nanofillers are easily aggregated in PP, because PP is a non-polar polymer and the interfacial connections between PP and nanofillers are weak. To construct a network system in PP, a SiO_2 precursor is impregnated in PP using supercritical carbon dioxide (CO_2), and nanofillers are subsequently formed by the sol-gel method [6]. As supercritical CO_2 can disperse the SiO_2 precursor into the amorphous region of PP and the sol-gel process can produce isolated nanofillers, this method is considered to be effective for producing PP/SiO_2 nanocomposites. TMOS, TEOS, or tetrapropoxysilane (TPOS) is used as the SiO_2 precursor. Random PP (molecular weight: 160,000, ethylene content: 2.5%), which has lower crystallinity than PP, is used as a matrix to increase the content of the impregnated SiO_2 precursor. A PP pellet is hot-pressed to obtain a PP film (thickness: 0.1 mm). PP/SiO_2 nanocomposites are prepared by the impregnation of the SiO_2 precursor in the amorphous PP using supercritical CO_2, which is followed by the sol-gel reaction of the precursor in the matrix using an acid catalyst.

4.1.3 What Kind of Mechanisms Are Working to Obtain Various Characteristics?

As the various characteristics of polymer nanocomposites are described in other chapters in detail, the mechanical, thermal, and electrical properties of polymer nanocomposites are only briefly introduced in this section.

Polymer nanocomposites have superior mechanical properties to the original polymer [5–8]. For example, in the case of polyimide (PI)/SiO$_2$ nanocomposites, the tensile strength and elongation at breaking increase with the SiO$_2$ content and reach a maximum at a SiO$_2$ content of 3 wt%, above which they decrease with the SiO$_2$ content [7]. The possible reasons for this are as follows. For a SiO$_2$ content of less than 3 wt%, the SiO$_2$ nanofillers are dispersed finely in the PI matrix and the interface between the PI and SiO$_2$ contains strong bonds. However, when the SiO$_2$ content is higher, e.g., 30 wt%, a sufficient number of SiO$_2$ nanofillers are aggregated for the interfacial connections between PI and SiO$_2$ to become weak.

Polymer nanocomposites also have higher heat-resistivity than the original polymer [4, 8]. For example, TEOS-PDMS nanocomposites have the property shown in Fig. 4.4 [4]. The figure shows the weight loss after aging at 200°C for 480 h as a function of the mole ratio of TEOS to PDMS. The decrease in the weight loss with increasing TEOS content may be due to the promotion of degradation and condensation polymerization reactions. Therefore, it is considered that the introduction of TEOS to PDMS suppresses the vaporization of free PDMS chains.

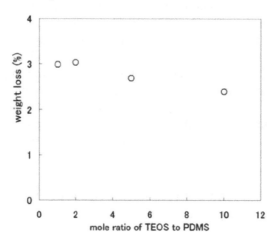

Figure 4.4 Change of weight loss of the TEOS-PDMS nanocomposites after isothermal aging at 200°C for 480 h in air [4].

Polymer nanocomposites also have superior electrical-insulating properties to the original polymer [4, 9]. For example, TEOS-PDMS nanocomposites have the property shown in Fig. 4.5 [4]. The figure shows the electrical resistivity before and after aging at 200°C for 480 h as a function of the mole ratio of TEOS to PDMS. After aging, the electrical resistivity is higher than that before aging, and the electrical resistivity of the nanocomposites is higher than that of PDMS. This may be due to the decrease in the ion carrier concentration caused by the reaction of TEOS and PDMS at a high temperature.

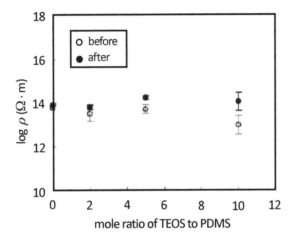

Figure 4.5 Volume electrical resistivity of the TEOS-PDMS nanocomposites before and after isothermal aging at 200°C for 480 h in air [4].

4.1.4 Composites Produced by the Sol-Gel Method Are Used in Daily Life

Although it used to be difficult to prepare thick-film products using the sol-gel method, some novel manufacturing methods have been developed recently. Table 4.2 shows some examples of commercialized polymer nanocomposites. Acryl, polyurethane, epoxy, and polyimide are used as matrix polymers, and the polymer nanocomposites are applied as functional coatings, adhesive agents, and electronic materials.

Table 4.2 Commercialized polymer nanocomposites produced by the sol-gel method [10]

Company	Structure	Production method	Examples of application	Notes
JSR Corp.	(a) Ceramic type (side-chain modification type) alkylalkoxysilane condensate (b) Polymer hybrid type acrylic polymer/silica	(a) Sol-gel reaction of alkyltrialkoxysilane (b) Sol-gel reaction of alkoxysilane and acrylic polymer containing alkoxysilyl group	Coatings for exterior building materials	Industrialized in 1993; product name "GLASCA"; awarded the Chemical Technology Prize of the Chemical Society of Japan in 1996
Arakawa Chemical Industries, Ltd.	Polyurethane, Epoxy resin, Polyamideimide, Polyimide/ Polyalkoxy-polysiloxane condensate	Sol-gel reaction of polymer and alkoxypolysiloxane oligomer containing functional group	Adhesive agents, functional coatings, electronic materials	Industrialized in 2000; product name "COMPOCERAN"
Hanse Chemie GmbH	Epoxy resin, etc./silica nanofillers (less than 50 wt%)	Synthesis of silica nanofillers in polymer by sol-gel reaction of sodium-silicate aqueous solution	Electric and electronic materials	Industrialized in 2001

References

1. Nagakura, S., Iguchi, H., Esawa, H., Iwamura, S., Sato, F., and Kubo, R. (1999). *Iwanami Physics and Chemistry Dictionary*, 5th ed. (Iwanami Shoten, Japan) (in Japanese).

2. Investigating R&D Committee on Technology and Application of Polymer Nanocomposites as Dielectric and Electrical Insulation (2006). *Technology and Application of Polymer Nanocomposites as Dielectric and Electrical Insulation*, Technical Report of IEEJ, No. 1051, p. 13 (in Japanese).

3. Fukuda, T., Fujiwara, T., Fujita, H., and Goda, H. (2005). Characteristics of Organic/Inorganic Nano-Hybrid Prepared by Site-Selectively Molecular Hybrid Method, *Seikei-Kakou*, **17**(2), pp. 109–205 (in Japanese).

4. Okamoto, T., Imasato, F., Shindo, T., and Nakamura, S. (2006). Electrical Insulating and Heat Resistive Properties Of TEOS-PDMS Hybrid Materials, *Proceedings of the 37th Symposium of the Electric and Electronic Insulating Material System*, pp. 117–120 (in Japanese).

5. Nakamura, S., Imasato, F., Kanamori, A., and Shindo, T. (2007). Influence of Gelation Conditions on Mechanical and Heat Resistive Properties of PDMS-TEOS hybrid material, *2007 National Convention Record IEEJ*, **2**(2–015), p. 16 (in Japanese).

6. Takeuchi, K., Terano, M., and Taniike, T. (2014). Sol-Gel Synthesis of Nano-Sized Silica in Confined Amorphous Space of Polypropylene: Impact of Nano-Level Structures of Silica on Physical Properties of Resultant Nanocomposites, *Polymer*, **55**, pp. 1940–1947.

7. Zhang, M. Y., Niu, Y., Chen, A. Y., Chang, W., and Hao, G. W. (2007). Study on Mechanical and Corona-Resistance Property of Polyimide/Silica Hybrid Films, *Proceedings of the International Conference on Solid Dielectrics*, pp. 353–356.

8. Liu, L., Weng, L., Zhu, X., Yang, L., and Lei, Q. (2009). The Effect of SiO_2/Al_2O_3 Weight Ratio on the Morphological and Properties of Polyimide/SiO_2-Al_2O_3 ternary hybrid films, *Proceedings of the 9th International Conference on Properties and Applications of Dielectric Materials*, pp. 777–780.

9. Shindo, T., Hishida, M., Sugiura, M., Nakamura, S., and Kamiya, K. (2004). Electrical Properties of Organic-Inorganic Hybrid Films Prepared by Sol-Gel Method (III), *2004 National Convention Record IEEJ*, **2**(2–082), p. 91 (in Japanese).

10. Chujo, K. (2003). Recent Situation of Polymer Nanocomposites Prepared by the Sol-Gel Method. *Plastics*, **54**(2), pp. 71–78 (in Japanese).

4.2 Mixing Methods for Quasi-Spherical Fillers (Thermoplastic and Thermosetting Resins)

Yuki Honda

Hitachi Metals, Ltd., Hitachi-shi, Ibaraki 319-1411, Japan

Polymers are broadly classified into thermoplastic resins, which melt at high temperatures, and thermosetting resins, which react by hardening without melting at high temperatures. A polymer nanocomposite is one of various polymers made with inorganic materials with diameters in the nano range containing a dispersed quasi-spherical filler made from an inorganic compound such as a metallic oxide. Polymer nanocomposites are new materials that have characteristics of both organic compounds and inorganic compounds. They may be used in various fields in the future. Extensive research of polymer nanocomposites is currently under way. This section describes methods of mixing a quasi-spherical nanofiller into a polymer as one way of creating a polymer nanocomposite.

4.2.1 Quasi-Spherical Nanofillers to Be Used Are Extremely Fine

A quasi-spherical filler consists of small inorganic particles with a diameter of 1–100 nm. The particles are composed of a metallic oxide, nitride, or carbide, etc., with typical examples, including silica (SiO_2), titania (TiO_2), zirconia (ZrO_2), aluminum oxide (Al_2O_3), boehmite aluminum oxide (AlOOH), barium titanate ($BaTiO_3$), boron nitride (BN), silicon carbide (SiC), and magnesium oxide (MgO).

In the past, a granular filler was made in a solid phase method in which bulk material was crushed; however, the diameters of particles in such granular fillers are over 100 nm. With advances in research, nanofiller particles with a diameter of less than 100 nm have been synthesized by liquid phase methods in which the nano-sized particles are made in a solution state, by refrigerating high-temperature vapor, or by a gas phase method with vapor phase reactivity.

4.2.2 Various Resins Are Used for Polymer Nanocomposites

Thermoplastic resins and thermosetting resins (or non-thermosetting resins) are used as the base resin for polymer nanocomposites. Base thermoplastic resins include polyethylene (PE), polypropylene (PP), polyamide (PA), ethylene-vinylacetate (EVA) etc. Base thermosetting resins (or non-thermosetting resins) include epoxy (EP), cross-linked polyethylene (XLPE), silicone rubber (SR), polyesterimide (PEI), polyamide-imide (PAI), polyimide (PI), etc. In particular, epoxy (EP) is used as the base polymer in many cases.

4.2.3 Polymer Nanocomposites Can Be Derived by Dispersing a Quasi-Spherical Nanofiller into a Polymer

A polymer nanocomposite can be derived directly by dispersing a nanofiller (that is, a filler of nano-sized quasi-spherical particles) into a polymer. However, mixing of a polymer and a granular nanofiller can cause strong agglomeration by cohesion among the filler particles, which results in a poor dispersion state. Therefore, to increase the affinity between nanoparticles and the polymer and to avoid agglomeration among the nanoparticles, the nanoparticles have surfaces improved by a silane coupling agent and a surface preparation agent.

For dispersion, in many cases, a high shearing force must be provided mechanically. Various dispersion devices are used: for example, high-pressure homogenizers, ultrasonic wave homogenizers, planetary stirring methods, and double spindle screw extruders.

The following subsections give examples of methods used to make a polymer nanocomposite.

4.2.3.1 Dispersing a quasi-spherical nanofiller in a thermosetting resin

Hydrophilic fumed silica with a particle diameter of 12 nm is prepared using a dry process (flame hydrolysis). The hydrophilic fumed silica is then dispersed in epoxy resin. Figure 4.6 shows a SEM (scanning electron microscope) photo of an epoxy/silica nanocomposite, in which silica filler particles that have diameters of under 20 nm are dispersed uniformly [1].

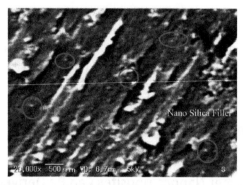

Figure 4.6   SEM photo of an epoxy/silica nanocomposite [1].

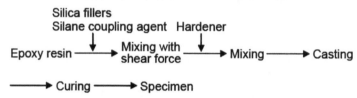

Figure 4.7   Method of preparing an epoxy/silica nanocomposite [2].

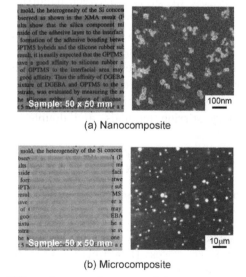

Figure 4.8   SEM photo of Epoxy/Silica nanocomposite and microcomposite [2].

Two epoxy/silica dispersions are compared below: a nanocomposite in which a silica filler with particle diameters of 12 nm is dispersed, and a microcomposite in which a silica filler with particle diameter of 1.6 nm is dispersed. The nanocomposite was prepared by the method described in Fig. 4.7. Figure 4.8 [2] shows the epoxy/silica nanocomposite. In the microcomposite, the silica particle diameters are at the µm level and the film is not transparent at 1 mm thickness. In the nanocomposite, the silica particle diameters are in the range of 20–80 nm, and dispersion is good with the film being transparent.

4.2.3.2 Dispersing a quasi-spherical nanofiller in a thermoplastic resin

This subsection reports on a polymer nanocomposite as a dispersed inflation film with magnesium oxide 5 phr (parts per hundred resin) uniformly dispersed in low-density polyethylene (LDPE). This is a nanocomposite with a quasi-spherical nanofiller dispersed in a thermoplastic resin. Figure 4.9 shows a TEM (transmission electron microscope) photo. An MgO nanofiller with a particle diameter of less than 200 nm is dispersed uniformly in LDPE [3].

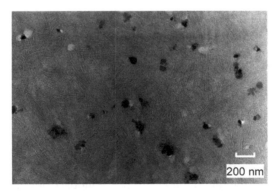

Figure 4.9 TEM photo of an LDPE/MgO nanocomposite [3].

4.2.3.3 Making a polymer nanocomposite from colloidal silica

Usually, a polymer nanocomposite is prepared by mixing a polymer and a quasi-spherical nanofiller. However, a new method is suggested, in which the polymer nanocomposite is made from colloidal silica. Silica sol, with an average particle diameter of 20 nm is mixed into polyesterimide or polyamide-imide and is dispersed with relatively weak agitation.

With the usual method of direct dispersion, nanoparticle cohesion occurs and particles form agglomerates of micro (μm) size. As shown in the TEM photo of Fig. 4.10 [4, 5], however, preparing the nanocomposite from colloidal silica makes a polymer nanocomposite without formation of agglomerates. As a result of this process, an insulation film for use on enameled wire can be obtained.

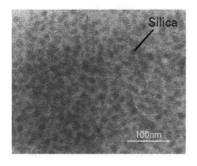 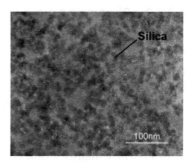

(a) Polyesterimide/silica   (b) Polyamide-imide/silica

Figure 4.10 TEM photo of a polymer nanocomposite made from colloidal silica [4, 5].

4.2.3.4 Nanoparticles can be dispersed by ultrasonic waves and centrifugal force

Recently, ultrasonic waves and centrifugal force have been used in a new nanoparticle dispersion technology as shown in Fig. 4.11 [6]. This method has resulted in lower agglomeration and has produced materials with excellent characteristics.

The agglomeration between the nanoparticles is disrupted by ultrasonic waves and the remaining agglomeration is almost completely removed by centrifugation, resulting in excellent dispersion. The base resin is epoxy, and the filling is globular alumina particles with an average diameter of 31 nm. The SEM photo of Fig. 4.12 [6] shows a polymer nanocomposite with a maximum diameter of agglomerated particles of 180 nm.

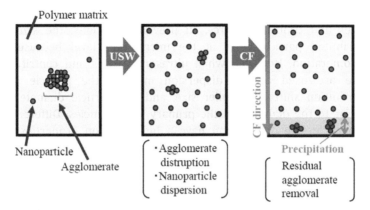

Figure 4.11 Schematic illustration of a particle dispersion effect by ultrasonic waves (USW) and centrifugal force (CF) (epoxy/alumina) [6].

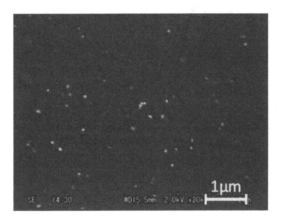

Figure 4.12 SEM photo of epoxy/alumina nanocomposite [6].

4.2.4 Control of Nanofiller Diameter Size Is Important for Good Nanocomposites

Even when nanofiller with small-diameter particles is used, agglomeration occurs, which can cause large decreases in material properties. Because of the very strong cohesion between quasi-spherical nanofiller particles, there is a need to control and suppress such agglomeration.

Figure 4.13 [7] illustrates improved dispersion by ultrasonic waves and centrifugal force. In previous methods, the size of distributed particles was at the micrometer level because of agglomeration. However, when ultrasonic waves and centrifugal force are used as the dispersion method, the particle size distribution shifts to smaller size, and the particle diameter is similar to the diameter of the primary nanoparticles. Differences in particle size distribution have a big effect on properties, so controlling diameter size is very important in obtaining stable characteristics of materials.

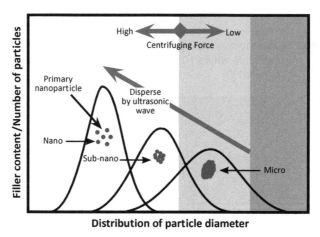

Figure 4.13 Illustration of using ultrasonic waves and centrifugal force to improve dispersion [7].

References

1. Iizuka, T., Uchida, K., and Tanaka, T. (2009). Voltage Endurance Characteristics of Epoxy/Silica Nanocomposites, *IEEJ Trans. Fundamentals Mater.*, **129**(3), pp. 123–127 (in Japanese).

2. Imai, T., Sawa, F., Ozaki, T., Shimizu, T., Kuge, S., Kozako, M., and Tanaka, T. (2006). Effects of Epoxy/Filler Interface on Properties of Nano- or Micro-composites, *IEEJ Trans. Fundamentals Mater.*, **126**(2), pp. 84–91.

3. Okuzumi, S., Masuda, S., Yoshinobu M., Masayuki N., Yoshinao M., Yoitsu S. (2007). *Electrical Breakdown Characteristic of MgO/LDPE Nanocomposite*, IEEJ, the 38th symposium on electrical and electronic insulating materials and applications in systems, F-3, pp. 141–146 (in Japanese).

4. Kikuchi, H., and Asano, K. (2006). Development of Organic/Inorganic Nano-composite Enameled Wire, *IEEJ Trans. Power Energy*, **126**(4), pp. 460–465 (in Japanese).

5. Kikuchi, H., Hanawa, H., and Honda, Y. (2012). Development of Polyamide-imide/Silica Nanocomposite Enameled Wire, *IEEJ Trans. Fundamentals Mater.*, **132**(3), pp. 263–269 (in Japanese).

6. Kurimoto, M., Watanabe, H., Hayakawa, N., Kato, K., and Okubo, H. (2009). Dispersibility and Dielectric Characteristics of Epoxy/Alumina Nanocomposite with Ultrasonic Wave and Centrifugal Force, *Proc. Tokai-Section Joint Conference on Electrical, Electronics, Information and Related Engineering*, O-409 (in Japanese).

7. Kurimoto, M., Watanabe, H., Kato, K., Hanai, M., Hoshina, Y., Takei, M., and Okubo, H. (2008). Dielectric Properties of Epoxy/Alumina Nanocomposite Influenced by Particle Dispersibility, *Annual Report, IEEE CEIDP*, No. 8-1, pp. 706–709.

4.3 Reactive Mixing Method for Fillers with Layered Structures

Takahiro Imai

Power and Industrial Systems R&D Center, Toshiba Corporation,
Fuchu-shi, Tokyo 183-8511, Japan
takahiro2.imai@toshiba.co.jp

Fillers with layered structures are generally called "clay." This "clay" is the same as the clay of "clay court" of tennis. Moreover, clay was used in the first nanocomposite in the world. Why is clay used in the fabrication of nanocomposites? Clay has a layered structure and contains metal ions between its layers. These ions can be exchanged for organic compounds that have an affinity to polymer. This property is essential for the fabrication of nanocomposites.

4.3.1 Unit Layer of Fillers with Layer-Structured Fillers Is 1 nm in Thickness

Clays, which comprise three-dimensional sheets with corners sharing SiO_4 tetrahedral sheets and $Al(OH)_6$ octahedral sheets, are hydrous aluminum phyllosilicates. They are classified into four typical groups: smectite, kaolin, chlorite, and illite. In particular, the smectite group, including montmorillonite, nontronite, and saponite, is suitable for reinforcement in clay nanocomposites. The structure of a clay (montmorillonite) is shown in Fig. 4.14 [1]. It has a laminated structure composed of a base unit (silicate layer) 100 nm long, 100 nm wide, and 1 nm thick. This layer comprises SiO_4 tetrahedral sheets and $Al(OH)_6$ octahedral sheets

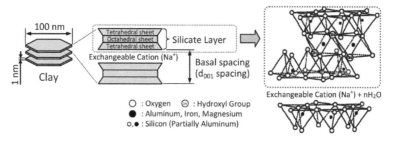

Figure 4.14 Structure of clay (montmorillonite) [1].

stacked in the order of tetrahedral sheet/octahedral sheet/ tetrahedral sheet. The properties of this clay (montmorillonite) are summarized in Table 4.3 [2].

Table 4.3 Properties of clay (montmorillonite) [2]

Properties		Unit	Value
Swelling		[ml/2 g]	65
pH (2% aqueous dispersion)		[—]	10.2
Electric conductivity		[μS/cm]	675
Viscosity (4% aqueous dispersion)		[mPa·s]	280
Visible light transmittance (1% aqueous dispersion)		[%]	1
Particle diameter		[nm]	100–2000
Specific surface (N_2, BET)		[m^2/g]	20
MB (methylene blue) absorbed amount		[mmol/100 g]	130
Cation exchange capacity		[meq/100 g]	108.6
Precipitation cation amount	Na^+	[meq/100 g]	114.1
	K^+	[meq/100 g]	2.8
	Mg^{2+}	[meq/100 g]	3.4
	Ca^{2+}	[meq/100 g]	18.2
Chemical composition	SiO_2	[%]	64.4
	Al_2O_3	[%]	25.9
	Fe_2O_3	[%]	3.5
	MgO	[%]	2.4
	CaO	[%]	0.7
	Na_2O	[%]	2.3
	K_2O	[%]	0.1

4.3.2 Organic Compounds Can Be Brought in between Neighboring Layers

In this clay, substitution of Al^{3+} atoms for Si^{4+} atoms in a tetrahedral sheet, or substitution of divalent metal ions for trivalent metal ions in an octahedral sheet, generates a negative charge. Cations such as sodium ions exist between the silicate layers to compensate for this negative charge. This structure imparts cation exchange ability to the clay. This characteristic of clay is represented by cation exchange capacity (CEC). In general, clay with CEC of approximately 100 meq/100 g is suitable for fabrication of clay nanocomposites. For example, montmorillonite has a CEC of 108.6 meq/100 g, as shown in Table 4.3.

The cation exchange reaction enables the clay to have an organic modifier (e.g. alkylammonium ion) between its silicate layers. The clay is thus organically modified by ion exchange between metal ions (sodium ions) and alkylammonium ions, as shown in Fig. 4.15. This procedure for organic modification is called "intercalation." Original clays with metal ions between clay layers have a hydrophilic property and lack affinity to the polymer. Therefore, organic modification is required for homogeneous dispersion in the polymer. Clays with alkylammonium ions between layers have a hydrophobic property, and have good affinity to the polymer.

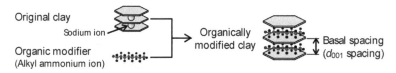

Figure 4.15 Organic modification of clay.

Alkylammonium ions are the most commonly used ions to modify clays. Typical alkylammonium ions are summarized in Table 4.4. In particular, quaternary alkylammonium ions are suitable as modifiers of clays for nanocomposites, and some kinds of clays modified with quaternary alkylammonium ions have been commercialized by clay manufacturers.

X-ray diffraction (XRD) analysis and thermal gravimetric analysis (TGA) provide information regarding the organic modification of clays. XRD measures the distance between silicate layers in the clays, which is called basal spacing (d_{001} spacing). Figure 4.16a compares the basal spacing of the original (unmodified) and the organically modified clay. The basal spacing of the modified clays is larger than that of the original clays, due to the insertion of alkylammonium ions instead of sodium ions between the clay layers. Moreover, Fig. 4.16b compares the thermal gravimetric curves of the original (unmodified) and the modified clay. The weight loss of the modified clay is larger that of the original clay. This is attributed to decreasing of the amount of organic modifier due to heating. Therefore, the rate of the organic modifier in the modified clay is estimated from the difference in weight loss in the TGA.

Table 4.4 Organic modifier of clays

Organic modifiers		Examples
Primary alkylammonium ion	H_2N^+-R (R: Alkyl group)	H_2N^+-$(CH_2)_{n-1}$-CH_3 (n = 8,1 1, 12,18 etc.) H_2N^+-$(CH_2)_{n-1}$-COOH (n = 8,1 1, 12,18 etc.)
Tertiary alkylammonium ion	R_1 | HN^+-R_2 | R_3 (R: Alkyl group)	CH_3 | HN^+-$(CH_2)_{11}$-CH_3 | CH_3 CH_3 | HN^+-$(CH_2)_{17}$-CH_3 | CH_3
Quaternary alkylammonium ion	R_1 | R_4-N^+-R_2 | R_3 (R: Alkyl group)	CH_3 | CH_3-N^+-$(CH_2)_{17}$-CH_3 | $(CH_2)_{15}$-CH_3 $(CH_2)_7$-CH_3 | CH_3-N^+-$(CH_2)_7$-CH_3 | $(CH_2)_7$-CH_3 T | CH_3-N^+-T | CH_3 $(CH_2)_2$-OH | CH_3-N^+-T | $(CH_2)_2$-OH CH_3 | CH_3-N^+-CH_2-⌬ | HT

(T: Tallow, HT: Hydrogenated tallow)

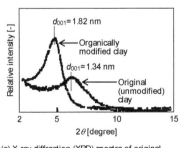

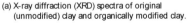

(a) X-ray diffraction (XRD) spectra of original (unmodified) clay and organically modified clay.

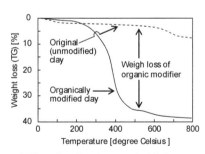

(b) Thermal gravimetric curves of original (unmodified) clay and organically modified clay.

Figure 4.16 Analysis of organically modified clay.

4.3.3 Layer-Structured Fillers Are Exfoliated and Dispersed

The first clay nanocomposite made in 1987 takes advantage of in situ polymerization [3]. The fabrication method for polyamide-based clay nanocomposite is shown in Fig. 4.17 [4]. In the first step, the original clays are organically modified by 12-amino-dodecanoic acid (H_2N-(CH_{11})-COOH) by a cation exchange reaction in an acidic solution. In the second step, melt mixing of the modified clay and ε-caprolactam (melting point 70°C) causes ε-caprolactam to be inserted in the interlayers. In the third step, ε-caprolactam is polymerized by a ring-opening reaction at 250°C. The basal spacing of the clays increases as polymerization progresses. Finally, the polyamide-based clay nanocomposite, in which the exfoliated clays are dispersed, is obtained.

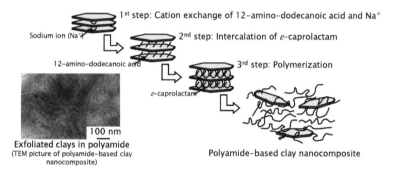

Figure 4.17 Fabrication method of polyamide-based clay nanocomposite [4].

An epoxy-based nanocomposite containing the clays is also made. The clays modified with dimethyl-tallow-alkylammonium ion (refer to Table 4.4) were dispersed in the acetone, and this clay/acetone mixture was mixed with the epoxy resin. After that, the acetone was removed by vacuum heating. The epoxy resin containing clays was cured in the presence of polydiamine hardener, and the epoxy-based nanocomposite was obtained. The clays are fully exfoliated in the nanocomposite as shown in Fig. 4.18 [5]. Moreover, various nanocomposites containing fully exfoliated clays were fabricated using a matrix polymer such as polystyrene, polypropylene, polylactic resin or acrylic resin [6–9].

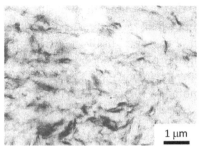 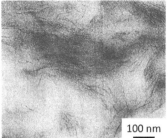

(a) Low magnification picture (b) High magnification picture

Figure 4.18 TEM pictures of fully exfoliated clays in epoxy-based nanocomposites [5].

4.3.4 State of Filler Dispersion Is Affected by Various Factors

Many factors involving the materials and exfoliation method affect the exfoliation and dispersion of clay in the fabrication of nanocomposites. The major factors are summarized in Table 4.5. It is important to fabricate clay nanocomposites with careful consideration of these factors, because exfoliation and homogenous dispersion can improve the properties of nanocomposites. There are two fabrication methods for clay nanocomposites: polymerization and melt blending, as shown in Fig. 4.19.

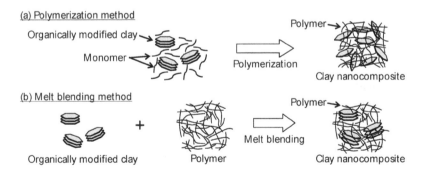

Figure 4.19 Exfoliation methods for organically modified clays in polymers.

As described for polyamide-based clay nanocomposite, organically modified clays are dispersed in the monomer in the polymerization method. After that, clay nanocomposites are obtained by polymerization. The polymerization reaction becomes a driving force for the exfoliation of the clays. In the melt blending method, the clay nanocomposites are obtained by blending of organically modified clays and the melt polymer. The dispersion or exfoliation of the clays is promoted by blending with mechanical shear.

Table 4.5 Major factor that influences exfoliation and dispersion of clays

Category		Major factor
Material	Monomer	–Chemical structure, etc.
	Polymer	–Chemical structure –Molecular weight –Polarity –Functional group, etc.
	Clay	–Shape –Size –Aspect ratio –Cation exchange capacity (CEC) –Purity, etc.
	Organic modifier (alkylammonium ions)	–Kinds of alkylammonium ions (Primary, tertiary and quaternary) –Length and branch of alkyl chain –Rate of the organic modifier in clay (Rate of residual Na^+), etc.
	Swelling solvent	–Polarity –Solubility parameter, etc.
Exfoliation method	Polymerization procedure	–Polymerization catalyst –Kinds of hardener, etc.
	Melt blending procedure	–Mixing equipment –Mixing time –Mixing temperature, etc.

Generally, the polymerization method requires a chemical plant to polymerize the monomer dispersed with organically modified clays, while the melt blending method has the advantage that the clay nanocomposites can be made relatively easily using thermoplastic resins. However, it is more difficult to make the clays fully exfoliate in the melt blending method. The mechanical properties and thermal resistance of polyamide-based clay nanocomposites made by these methods are summarized in Table 4.6 [2]. This table demonstrates that polyamide-based clay nanocomposite made by the polymerization method has superior flexural strength, flexural modulus and heat distortion temperature (HDT) to those made by the melt blending method. These results are attributed to the difference between the exfoliations of the clays in the polyamide.

Table 4.6 Comparison of polymerization method and melt blending method in polyamide-based clay nanocomposites [2]

Properties		Polyamide[a] (PA6)	PA6/clay nanocomposites	
			Polymerization method	Melt blending method
Amount of clays (Nanomer 1.24TC[b])	[wt%]	0	5.5	5.5
Flexural strength	[MPa]	97.5	143.3	124.3
Flexural modulus	[MPa]	2420	4247	3740
Heat distortion temperature (HDT, 264 psi)	[°C]	59.8	131.9	116.4

[a]Capron 8202 35FAV manufactured by Allied Sigma company.
[b]Organically modified clay manufactured by Nanocor company.

Furthermore, the influence of the kinds of amine hardener and the mixing time of epoxy resin and organically modified clays on the exfoliation of clays is reported for epoxy-based nanocomposites.

Figure 4.20a shows the influence of hardener (curing agent) reactivity on the structure of the nanocomposites [10]. In nanocomposites including organically modified clays with octadecylammonium ion (H_2N^+-$(CH_2)_{17}$-CH_3), each basal spacing (d_{001} spacing) is dependent on the kind of hardener. Clay anocomposite cured with (i) PACM had a diffraction peak attributed to 3.7 nm of basal spacing. However, the basal spacing of clay nanocomposite cured with (ii) DDDHM expanded to 4.0 nm. This indicates that the clays were more exfoliated in the nanocomposite cured with DDDHM than in the nanocomposite cured with PACM.

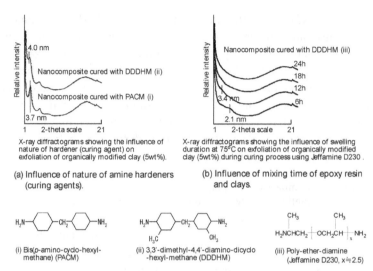

Figure 4.20 Influence of kinds of amine hardener and mixing time of epoxy resin and clays on exfoliation of clays in epoxy-based nanocomposites [10].

In addition, Fig. 4.20b shows the influence of mixing time of the epoxy resin and clays on the structure of the nanocomposites. The epoxy resin and the clays modified by octadecylammonium ions were mixed for 6, 12, 18, and 24 h. Each mixture was then cured with (iii) poly-ether-diamine (Jeffamine D230). The WAXD spectra of the nanocomposites show that the basal spacing of the clay expanded from 2.1 to 3.4 nm with increase in mixing time. The diffraction peak, which is attributed to the laminated structure of the clay, disappeared from the nanocomposite after 24 h of mixing. This nanocomposite had fully exfoliated clays.

4.3.5 Various Techniques Are Developed for Homogeneous Dispersion

Exfoliation and dispersion methods for clays are ever improving. A brief introduction to the four specific exfoliation and dispersion methods is presented here.

4.3.5.1 Dispersion of clays swelled with organic solvent

Figure 4.21 shows methods using swelled clays and a curing accelerator. Organically modified clays have a swelling characteristic.

An organic solvent (polar solvent) such as *N,N*-dimethylacetamide (DMAc) is inserted into the galley between the clay layers and expands the basal spacing of the clays [11].

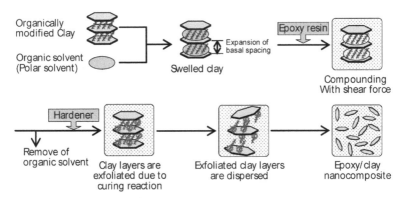

Figure 4.21 Method using swelled clays and curing accelerator.

Generally, the curing reaction between the epoxy resin and the acid anhydride hardener is slow, and an accelerator is sometimes used to speed it up. The tertiary ammonium ion, which is one of the accelerators, is inserted between the clay layers. The clays modified by the tertiary ammonium ions are dispersed in the epoxy resin. The clay nanocomposite is then obtained by curing in the presence of the hardener. In this system, the curing reaction starts between the clay layers in which the accelerators exist. Therefore, the cross-linking network of the epoxy resin starts to form from inside the clay layer. The beginning of the curing reaction between the clay layers promotes exfoliation of the clays.

4.3.5.2 Clay dispersion using AC voltage application

Figure 4.22 shows the method in which AC voltage is applied [12]. The 1 phr (parts per hundred parts of resin) clays modified with dimethyl-benzyl-hydrogenated tallow quaternary ammonium ions are dispersed in the epoxy resin. After that, AC voltage application using parallel electrodes with a 40 mm gap promotes the exfoliation of the clays. The hopping current increases rapidly, and the exfoliation of the clays occurs in process (II). The exfoliated clays are dispersed in the epoxy resin in process (III).

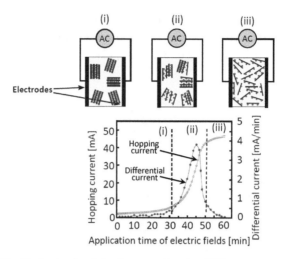

Figure 4.22 Method using AC voltage application [12].

4.3.5.3 Clay dispersion using high-shear compounding in solid phase

Figure 4.23 shows the methods using high-shear compounding in solid phase [13]. When the shear rate increases in the compounding process, the polymer containing the clays tends to show non-Newtonian behavior due to the decreasing of the polymer viscosity. Therefore, high-shear compounding in the solid phase using an extruder is effective in dispersing the clays homogenously.

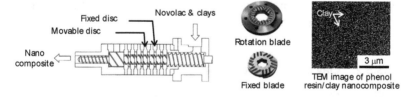

Figure 4.23 Method using high-shear compounding in solid state [13].

Novolac (M_n = 1600, M_w = 7400) containing 4 phr organically modified clays in solid phase are mixed with high shear force in an extruder. The fourth compounding (passing through the extruder) disperses the clays homogenously. After that, phenol resin/clay nanocomposite is obtained by curing in the presence of a hardener.

4.3.5.4 Dispersion of clays without organic modification

Figure 4.24 shows the method using unmodified clays [14]. Organic modification of clays has been considered an essential process in the fabrication of nanocomposites. However, a low-cost fabrication method using unmodified clays has been developed.

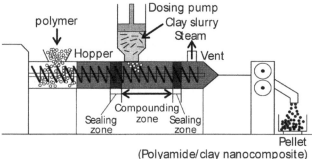

Figure 4.24 Method using unmodified clays [14].

The water slurry technique disperses unmodified clays in polyamide. Polyamide resin is supplied from the hopper of the extruder, and the clay slurry is injected from the dosing pump. The clays exfoliate during the compounding process, and steam is exhausted from the vent at the end of the extruder. The polyamide/clay nanocomposite is finally processed into pellets.This nanocomposite using unmodified clays is inferior to that using organically modified clays. However, this fabrication method without organic modification of clays seems to be an innovative method from the view-point of industrial manufacturing.

As mentioned above, this section has described the characteristics of fillers with layered structures (clays) and the fabrication method for clay nanocomposites. Homogenous dispersion of clays that are organically modified using the ion exchange reaction (intercalation) is one of the most important methods for nanocomposite fabrication.

References

1. Imai, T., Sawa, F., Ozaki, T., and Shimizu, T. (2004). Partial Discharge Resistance Enhancement of Organic Insulating Materials by Nano-Size Particle Dispersion Technique, *IEEJ Paper of Technical Meeting*

on Dielectrics and Electrical Insulation, No. DEI-04-78, pp. 35–38 (in Japanese).

2. Chujo, K., and Abe, K. (2001). *Technological Trend of Polymer Nano-Composites*, CMC Publishing Co., Ltd., p. 17, 46, 48 (in Japanese).

3. Fukushima, Y., and Inagaki, S. (1987). Synthesis of an Intercalated Compound of Montmorillonite and 6-Polyamide, *J. Inclusion Phenom.*, **5**, pp. 473–482.

4. Kato, M. (2009). *Engineering Materials*, **57**(7), pp. 31–35 (in Japanese).

5. Brown, J. M., Curliss, D., and Vaia, R. A. (200). Thermoset-Layered Silicate Nanocomposites. Quaternary Ammonium Montmorillonite with Primary Diamine Cured Epoxies, *Am. Chem. Soc. Chem. Mater.*, **12**(11), pp. 3376–3384.

6. Zhu, J., Morgan, A., Lamelas, F., and Wilkie, C. (2001). Fire Properties of Polystyrene-Clay Nanocomposites, *Am. Chem. Soc. Chem. Mater.*, **13**(10), pp. 3774–3780.

7. Okamoto, M., Nam, P., Maiti, P., Kotaka, T., Nakayama, T., Takada, M., Ohshima, M., Usuki, A., Hasegawa, N., and Okamoto, H. (2001). Biaxial Flow-Induced Alignment of Silicate Layers in Polypropylene/Clay Nanocomposite Foam, *Am. Chem. Soc. Nano Letters*, **1**(9), pp. 503–505.

8. Krikorian, V., and Pochan, D. (2003). Poly (L-Lactic Acid)/Layered Silicate Nanocomposite: Fabrication, Characterization and Properties, *Am. Chem. Soc. Chem. Mater.*, **15**(22), pp. 4317–4324.

9. Yeh, J., Liou, S., Lin, C., Cheng, C., and Chang, C. (2002). Anticorrosively Enhanced PMMA-Clay Nanocomposite Materials with Quaternary Alkylphosphonium Salt as an Intercalating Agent, *Am. Chem. Soc. Chem. Mater.*, **14**(1), pp. 154–161.

10. Kornmann, X., Lindberg, H., and Berglund, L. A. (2001). Synthesis of Epoxy–Clay Nanocomposites. Influence of the Nature of the Curing Agent on Structure, *Polymer*, **42**, pp. 4493–4499.

11. Harada, M., Aoki, M., and Ochi, M. (2008). *J Network Polymer, Japan*, **29**(1), pp. 38–43 (in Japanese).

12. Park, J., and Lee, J. (2010). A New Dispersion Method for the Preparation of Polymer/Organoclay Nanocomposite in the Electric Fields, *IEEE Trans. Dielectr. Electr. Insul.*, **17**, pp. 1516–1522.

13. Matsumoto, A., Otsuka, K., Kimura, H., and Yamaoka, K. (2009). *Proceedings of the 59th the Network Polymer Symposium Japan*, pp. 233–234 (in Japanese).

14. Hasegawa, N., and Usuki, A. (2001). Development of polyamide-clay nanocomposites, *Proceedings of the Annual Meeting of the Japan Society of Polymer Processing (JSPP)*, pp. 153–154 (in Japanese).

4.4 Surface Modification of Nanofiller Helps Uniform Dispersion

Minoru Okashita

SWCC Showa Cable Systems Co., LTD.,
Sagamihara-shi, Kanagawa 252-0253, Japan

m.okashita202@cs.swcc.co.jp

Many cases have been reported that the dispersion of a nanofiller (nanoparticles) in a resin improves the characteristics of the resin, but the small particle size tends to induce secondary agglomeration, which may rather hamper dispersion of the particles in a resin, resulting contrarily in poorer resin characteristics. This effect can be avoided by the modification of the nanoparticle surface, thus permitting the manufacture of nanocomposite materials with improved characteristics. This section discusses the surface modification of nanoparticles and its effects.

4.4.1 Surface Modification Is Important

One of fundamental materials in nanotechnology, nanoparticles are generally particles with diameters of 0.1 μm (100 nm) or smaller. Recent trends in customer needs towards higher performance and function, miniaturization and resource of equipment and electronic devices saving have prompted improvements of characteristics of conventional insulation materials.

Last few years witnessed emerging investigations on nanocomposite materials consisting of nanoparticles dispersed in a variety of resin matrices, of which mechanical, electric, thermal, optical and chemical properties have extensively been reported [1].

Improvements in these properties owe largely to the very large contact surface area of the nanoparticles of very small diameters in apparent size or steric hindrance that increases electron or photon path length due to greater particle counts in the unit volume of the resin matrix. It is also possible to improve mechanical properties without sacrificing transparency by using nanoparticles of diameters smaller than the wavelength of visible light (1–10 nm).

On the other hand, small particle sizes mean high surface activity, which facilitates secondary coagulation of the particles into coarse lumps. This often presents serious problems in manufacturing nanoparticles and in mixing of nanoparticles and resin. A method to solve these problems is treatment of the nanoparticle surface with certain organic compounds to prevent secondary coagulation and enhance the affinity of the surface with the resin for easier dispersion.

Figure 4.25 shows transmission electron microscope (TEM) images of lamellar silicate (clay) particles dispersed in an epoxy resin with and without treatment with silane coupling; Fig. 4.26 shows TEMs of silica dispersed in a cycloaliphatic epoxy resin with and without treatment with silane coupling [2–3]. It is clear that the treated clay and silica particles are dispersed more uniformly than the untreated ones, showing less agglomeration.

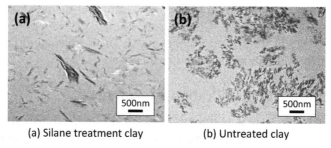

(a) Silane treatment clay   (b) Untreated clay

Figure 4.25 TEM images of epoxy/clay composites [2].

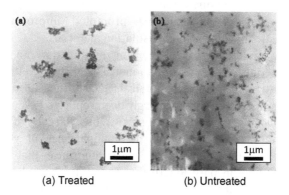

(a) Treated   (b) Untreated

Figure 4.26 TEM images of nanocomposites showing the dispersion states of the silica nanoparticles. © 2010 IEEE. Reprinted, with permission, from Huang, X., Zheng, Y., and Jiang, P. (2010). Influence of Nanoparticle Surface Treatment on the Electrical Properties of Cycloaliphatic Epoxy Nanocomposites, *IEEE Trans. Dielectr. Electr. Insul.*, **17**(2), pp. 635–643.

The surface modification of nanoparticle surface is thus a very important technique to obtain nanocomposite materials with intended characteristics.

4.4.2 Several Methods Are Available for Surface Modification

As stated above, surface modification is indispensable to mix nanoparticles with resin. A common method of surface modification is treatment of inorganic particles with a silane-coupling agent to enhance the affinity with the resin [3–4].

The reaction of the silane-coupling agent with silica substrate is shown in Fig. 4.27. The low affinity with organic reagents of silica surface can be enhanced by grafting an organosilane compound on it as illustrated.

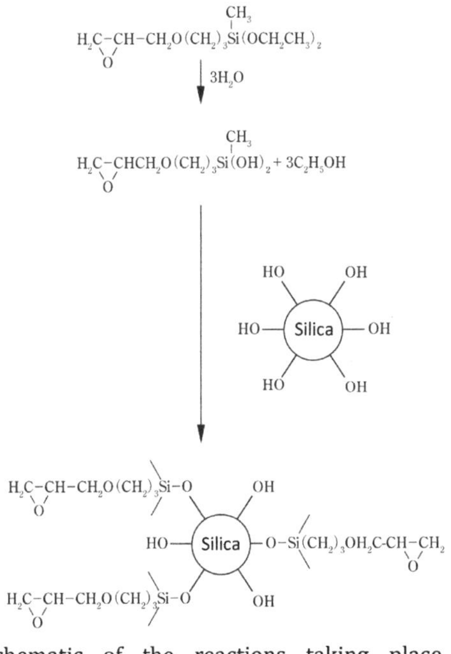

Figure 4.27 Schematic of the reactions taking place between the original silica and the silane. © 2010 IEEE. Reprinted, with permission, from Huang, X., Zheng, Y., and Jiang, P. (2010). Influence of Nanoparticle Surface Treatment on the Electrical Properties of Cycloaliphatic Epoxy Nanocomposites, *IEEE Trans. Dielectr. Electr. Insul.*, **17**(2), pp. 635–643.

Figure 4.28 shows the Weibull distribution of the water tree length in XLPE specimens 2 mm in thickness to which a voltage of 5 kV AC with a frequency of 1.5 kHz was applied in a 1 M NaCl solution at room temperature for 45 days. The specimens LDPE-200, LDPE-O and LDPE-D contain, respectively, surface-untreated silica nanoparticles (Aerosil 200), surface-treated Aerosil 200 with octyltrimethoxysilane, or surface-treated Aerosil 200 with dimethyldichlorosilane. The growth of water tree in the nanocomposite material depends on the surface conditions of the silica nanoparticles produced by different silane-coupling agents [4].

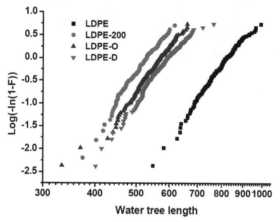

Figure 4.28 Weibull distributions of water tree lengths. © 2009 IEEE. Reprinted, with permission, from Huang, X., Ma, Z., Jiang, P., Kim, C., Liu, F., Wang, G., and Zhang, J. (2009). Influence of Silica Nanoparticle Surface Treatments on the Water Treeing Characteristics of Low Density Polyethylene, *Proceedings of the IEEE ICPADM*, No. H-7, pp. 757–760.

Many cases using surface modifiers other than silane-coupling agents are also reported. Table 4.7 summarizes combinations of the particles and treating agents along with their effects.

Polypropylene (PP)-based nanocomposites do not always acquire sufficient mechanical characteristics only by nanoparticle treatment with a silane-coupling agent; grafting of PP chains on the nanoparticle surface is proposed as an alternative [12–13].

Table 4.7 The kinds and effects of the treatment agent of nanoparticles

Composite polymer	Particles	Treatment agent	Effect	Ref.
Epoxy	Aluminum nitride Boron nitride	Polysaccharides hydrophilic group Silane-coupling agent	Thermal conductivity Anti-water treeing	[5–8]
Epoxy	Carbon nanotube Silicon carbide	Triethylenetetraamine Silane-coupling agent	Thermal conductivity	[9]
PVDF	Carbon nanotube	HNO_3/H_2SO_4	Conductivity Dielectric constant	[10]
EVA	Clay	Ammonium salt	Space charge	[11]

Figure 4.29 compares TEM images of silica and PP-grafted silica nanocomposite material dispersed in PP matrix; Fig. 4.30 shows the mechanical properties of the PP/PP-grafted silica nanocomposite material. It is clear that the PP-grafted silica gives better mechanical properties, while showing largely the same dispersion in both cases.

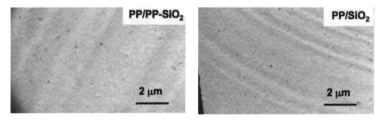

Figure 4.29 TEM image of the nanocomposites [12].

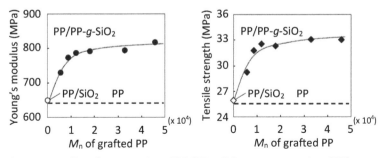

Figure 4.30 Tensile properties of PP/PP-g-SiO$_2$ nanocomposites [13].

A study of treatment of carbon nano-balls (CNBs) with solution plasma (plasma generated in a liquid) found that CNB beads with a diameter of 800 nm (Fig. 4.31) can be dispersed uniformly in water by hydrophilizing by solution plasma treatment (Fig. 4.32), and gives a nylon-6 composite material with improved mechanical properties [14].

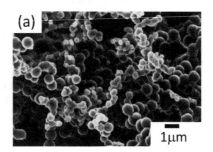

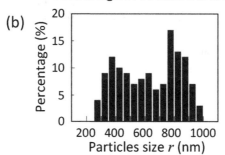

Figure 4.31 Carbon nano-balls [14].

(a) Before solution plasma modification

(b) After solution plasma modification.

Figure 4.32 Photographs of the solutions added CNBs [14].

4.4.3 Larger Filler Particles May Be Surface-Modified by Nanofillers

Nanocomposite materials employing electrostatic adsorption as a new surface modification technique has been reported [15]: composite particles were manufactured by making fine alumina particles adsorbed electrostatically onto poly(methyl methacrylate) particles (Fig. 4.33).

Figure 4.33 SEM image of PMMA-alumina particles [15].

It has also been attempted to coat electrically conductive particles with alumina nanoparticles to turn them into insulating ones [16]. Figure 4.34 shows a SEM image of spherical graphite powder coated with alumina nanoparticles. The image (b) of the nano-alumina-coated graphite clearly shows the alumina particles adhere to the surface of the core uniformly and densely.

 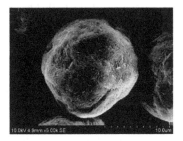

(a) Without coating   (b) With a nano-alumina coating

Figure 4.34 SEM images of graphite particles [16].

Figure 4.35 shows a TEM image of a spherical aluminum particle coated with alumina nanoparticles. This figure shows that an alumina film of 10–20 nm thick is formed on the surface of the core. Coating of various core materials with alumina nanoparticles has proved to be able to change conductive particles into insulating ones without affecting the thermal conductivity.

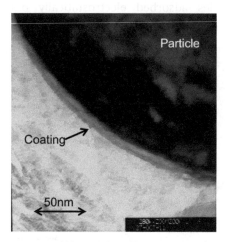

Figure 4.35   TEM image of Al particle with a nano-alumina coating [16].

References

1. Investigating R&D Committee on Technology and Application of Polymer Nanocomposites as Dielectric and Electrical Insulation. (2006). Technology and Application of Polymer Nanocomposites as Dielectric and Electrical Insulation, *IEEJ Technical Report*, No. 1051.

2. Yamazaki, K., Imai, T., Ozaki, T., Shimizu, T., Kurosawa, K., and Kitamura, H. (2010). Characteristics of Epoxy/Silane-treated Clay Nanocomposites, *Proceedings of the Technical Society Conference (Fundamentals and Materials)*, No. XV-9, p. 348.

3. Huang, X., Zheng, Y., and Jiang, P. (2010). Influence of Nanoparticle Surface Treatment on the Electrical Properties of Cycloaliphatic Epoxy Nanocomposites, *IEEE Trans. Dielectr. Electr. Insul.*, **17**(2), pp. 635–643.

4. Huang, X., Ma, Z., Jiang, P., Kim, C., Liu, F., Wang, G., and Zhang, J. (2009). Influence of Silica Nanoparticle Surface Treatments on the Water Treeing Characteristics of Low Density Polyethylene, *Proceedings of the IEEE ICPADM*, No. H-7, pp. 757–760.

5. Hirano, H., Hasegawa, K., Agari, Y., Ishikawa, H., and Nagayama, K. (2009). *Proceedings of the 59th Symposium on Network Polymer*, No. PO-26, pp. 269–270 (in Japanese).

6. Okazaki, Y., Ohki, Y., Tanaka, T., Kaneko, S., Okabe, N., and Kozako, M. (2008). *IEEJ Kyusyu Branch Preprints*, No. 05-2A-06 (in Japanese).

7. Hanagasaki, H., Ohashi, T., and Suenaga, H. (2006). Research on the Properties of Treated BN Filler as Material of Heat Releasing Resin, *Bull. West. Hiroshima Prefecture Ind. Res. Inst.*, No. 49–19, pp. 1–4.

8. Xu, Y., and Chung, D. D. L. (2000). Increasing the Thermal Conductivity of Boron Nitride and Aluminum Nitride Particle Epoxy-Matrix Composites by Particle Surface Treatments, Composite Interfaces, *Composite Interfaces*, **7**(4), pp. 243–256.

9. Yang, K., and Gu, M. Y. (2010). Enhanced Thermal Conductivity of Epoxy Nanocomposites Filled with Hybrid Filler System of Triethylenetetramine-functionalized Multi-walled Carbon Nanotube/ Silane-modified Nano-sized Silicon Carbide, *Composites Part A–Appl. Sci. Manufacturing*, **41**, pp. 215–221.

10. Dang, Z. (2006). High Dielectric Constant Percolative Nanocomposites Based on Ferroelectric Poly(vinylidene fluoride) and Acid-Treatment Multiwall Carbon Nanotubes, *Proceedings of the IEEE ICPADM*, No. P3-16, pp. 782–786.

11. Montanari, G. C., Cavallini, A., Guastavino, F., Coletti, G., Schifani, R., M. di Lorenzo del Casale, Camin, G., and Deorsola, F. (2004). Microscopic and Nanoscopic EVA Composite Investigation: Electrical Properties and Effect of Purification Treatment, *Ann. Rept. IEEE CEIDP*, No. 4–3, pp. 318–321.

12. Umemori, M., Taniike, T., Terano, M., Kawamura, T., and Nitta, K. (2009), Synthesis of Polypropylene-Based Nanocomposites Using Silica Nano Composites Chemically Modified with the Matrix Polymer Chains, *Polymer Preprints, Japan*, No. 2D06, p. 702.

13. Toyonaga, M., Umemori, M., Taniike, T., and Terano, M. (2012). Relationship of Grafted Chain and Reinforcements on Polypropylene/ Polypropylene Grafted Silica Nanocomposite, *Polym. Preprints, Japan*, No. 3Pc101, p. 1015.

14. Hieda, J., Shirafuji, T., Noguchi, Y., Saito, N., and Takai, O. (2009). Solution Plasma Surface Modification for Nanocarbon-Composite Materials, *J. Japan Inst. Metals*, **73**(12), pp. 938–942.

15. Hirota, S., Murakami, Y., Hakiri, N., Muto, H., Kurimoto, M., and Nagao, M. (2011). Experimental Production of Nanocomposite

Insulating Material Using Electrostatic Adsorption Method, *Procceedings of the Technical Society Conference, Fundamentals and Materials*, No. XIX-10, p. 409.

16. Sato, G., Sato, M., Nakamura, T., Kinoshita S., Kozako, M. and Hikita, M. (2011). Fundamental Examination on Insulation Performance of Conductive Fillers Using Nano-Alumina Hydrate Coating Technique, *IEEJ, The Paper of Technical Meeting on Dielectrics and Electrical Insulation*, No. DEI-11-087, pp. 59–64.

PART 3

FUNDAMENTALS
(MATERIAL CHARACTERISTICS)

Chapter 5

Drastic Improvement of Dielectric Performances by Nanocomposite Technology

Muneaki Kurimoto[a], Kazuyuki Tohyama[b], Yasuhiro Tanaka[c], Yoshinobu Mizutani[d], Toshikatsu Tanaka[e], Masayuki Nagao[f], Naoki Hayakawa[a], Takanori Kondo[g] and Tsukasa Ohta[h]

[a]Nagoya University
[b]National Institute of Technology, Numazu College
[c]Tokyo City University
[d]Central Research Institute of Electric Power Industry (CRIEPI)
[e]Waseda University
[f]Toyohashi University of Technology
[g]NGK INSULATORS, LTD.
[h]Mie University

5.1 Permittivity and Dielectric Loss: Dielectric Spectroscopy

Muneaki Kurimoto

Department of Electrical Engineering and Computer Science, Nagoya University, Nagoya-shi, Aichi 464-8603, Japan

kurimoto@nuee.nagoya-u.ac.jp

The changes in the movement and structure of atoms and molecules as a result of using nanocomposites are represented

---

Advanced Nanodielectrics: Fundamentals and Applications
Edited by Toshikatsu Tanaka and Takahiro Imai
Copyright © 2017 Pan Stanford Publishing Pte. Ltd.
ISBN 978-981-4745-02-4 (Hardcover), 978-1-315-23074-0 (eBook)
www.panstanford.com

by the dielectric properties of the nanocomposites. In particular, the permittivity of nanocomposites can increase or decrease depending on the condition, and can be used as an index to evaluate the properties of nanocomposites. The phenomenon of decreasing the permittivity of the polymer by adding nanofillers is unique to nanocomposites. In this section, the effects of nanofiller interfaces on the polymer and of the volume percentage of nanofillers on the nanocomposites are examined.

5.1.1 Permittivity and Dielectric Loss Are Evaluated by Dielectric Spectroscopy Dependent on Frequency and Temperature

Only negligible current flows when a voltage is applied to an electrical insulator; however, positive and negative charges constituting the atoms and molecules in the material are displaced because of the application of electric field to the atoms and molecules. When freely moving ions exist in the material, positive and negative ions are separated to induce the displacement of charges. The phenomenon is called dielectric polarization. For example, epoxy resin and water molecules have polarities and are align along the direction of the electric field against thermal motion, inducing an electric displacement in the material, as shown in Fig. 5.1.

The effect of dielectric polarization is represented by the capacitance of a parallel-plate capacitor with an insulating material placed between the plates (C). Assuming that the capacitance of a capacitor of the same size with a vacuum between the plates is C_0,

$$\frac{C}{C_0} = \varepsilon_r \tag{5.1}$$

ε_r is always ≥ 1 regardless of the insulating material placed between the plates. This is because dielectric polarization occurs in the material between the plates, leading to the storage of many charges and an increase in capacitance. ε_r is called the relative permittivity and is material-specific. Table 5.1 shows the relative permittivities of typical insulating materials. The permittivity ε is also used and calculated as the product of ε_r and the permittivity of vacuum ($\varepsilon_0 = 8.854 \times 10^{-12}$ F/m), namely, $\varepsilon_r = \varepsilon_0 \varepsilon_r$.

 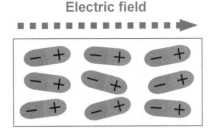

Figure 5.1 Electric displacement in the material (dipole polarization).

Table 5.1 Relative permittivity of typical insulating materials

Insulating material	Relative permittivity
Air	1.0003
Oil	2.2–2.4
Distilled water	81
Polyethylene	2.2–2.3
Epoxy resin	3.0–4.5
Silica	3.7–4.5
Magnesia	9–10
Alumina	9–10
Titania (anatase type crystal)	30–50
Titania (rutile type crystal)	90–120

There are several types of dielectric polarization, including electronic polarization, atomic polarization, dipole polarization, space-charge polarization, and interfacial polarization. These polarizations are defined as follows:

(i) Electronic polarization: the polarization caused by the displacement of atomic nuclei and electronic clouds in the direction of electric field.

(ii) Atomic polarization: the polarization observed when the distance between ion pairs changes in ionic crystals.

(iii) Dipole polarization: the polarization observed when molecules that originally have displaced charges (polar molecules) align along the direction of electric field against thermal motion.

(iv) Space-charge polarization: the polarization observed when a substance added to a material becomes free ions that accumulate near one of the electrodes as charges.

(v) Interfacial polarization: the polarization observed when the charges accumulate at the interface between different materials.

When a DC voltage is applied to an insulating material, all types of dielectric polarization appear and the relative permittivity increases. However, when an AC voltage is applied to an insulating material, the types of dielectric polarization that appear differ because the rates of the changes in the intensity and direction of electric field depend on frequency. The frequency characteristics of the permittivity and dielectric loss are shown in Fig. 5.2.

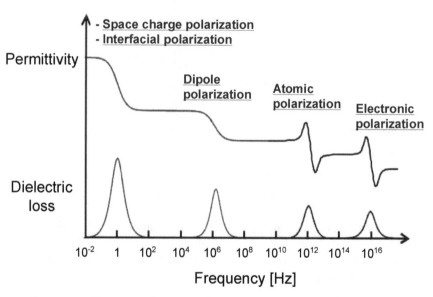

Figure 5.2 Frequency characteristics of permittivity and dielectric loss.

For general insulating materials, space-charge polarization and interfacial polarization do not appear at frequencies higher than several kHz; only dipole polarization, atomic polarization, and electronic polarization appear, leading to the decrease in permittivity. For higher frequencies such as those in the infrared or ultraviolet frequency range, only atomic polarization and electronic polarization appear. This means that, with increasing frequency or rates of changes in the intensity and direction of electric field, some types of polarization become too slow to follow the changes. This phenomenon is called dielectric dispersion. For example, in the case of dipole polarization, the time lag of polarization is caused by collisions between molecules and constraint from the surrounding molecules, and is consumed as thermal energy in the material. This power loss is called dielectric loss, which has an extreme value in a specific frequency range. The energy absorption in this frequency range is called dielectric absorption.

The frequency characteristics of the permittivity and dielectric loss are significantly affected by the movement and structure of atoms and molecules. Their temperature characteristics are also affected similarly. These behaviors are collectively represented by dielectric spectra.

How are dielectric spectra obtained? The method of obtaining dielectric spectra depends on the frequency range of the voltage used for the measurement. In general, the dielectric characteristics at $\leq 10^8$ Hz are obtained by measuring the AC impedance (capacitance and dielectric tangent) of specimens. As shown in Fig. 5.3, a plate or sheet of insulating material is used as a specimen, and is sandwiched with circular metal electrodes to form a capacitor. To avoid gaps between a specimen and electrodes, a metal is vapor-deposited or a metal paste is coated on the specimen. Although guard electrodes are generally used to avoid the effect of electric distortion at the edges of the capacitor, the guard electrodes may not be used if the electrode area is sufficiently large compared with the thickness of the specimen. A Schering bridge, an inductance, capacitance, and resistance (LCR) meter, and an impedance analyzer are used for measurement.

The dielectric properties in the range of 10^8 to 10^{11} Hz are determined by measuring the propagation characteristics (reflection and transmission) of electromagnetic waves. A network analyzer is sometimes used for measurement. The dielectric characteristics in higher frequency ranges are determined by obtaining the absorption spectra of electromagnetic waves and measuring the refractive index of the specimen. A Fourier-transform infrared (FT-IR) spectrometer and an ellipsometer are used for measurement.

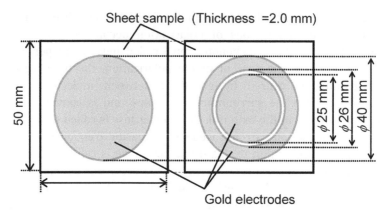

Figure 5.3 A plate or sheet of insulating material is used as a specimen, and is sandwiched with circular metal electrodes to form a capacitor.

5.1.2 Permittivity of Microcomposites Is Determined by Composition Ratio

How does the relative permittivity of a composite change upon the addition of micrometer-sized ceramic fillers to a polymer? The relative permittivity of the composite is roughly calculated using the relative permittivities of the polymer and ceramic and their composition ratio (volume percentage), which agrees fairly well with the experimental value. Strictly, it is also affected by the polarization state, shape, and distribution at the interface of materials and has been a target of research for many years. A general theory used to calculate the relative permittivity of the composite is introduced below. Here, a case in which a ceramic filler with a high relative permittivity is added to a polymer with a low relative permittivity is used as an example.

When interfacial polarization does not appear at the interface between the filler and the polymer, the relative permittivity of the composite can be calculated using a model that simulates the two materials as capacitors. The relative permittivity of the composite consisting of two materials is calculated by

$$\varepsilon_a^k = V_p \cdot \varepsilon_p^k + (1 - V_p) \cdot \varepsilon_m^k \tag{5.2}$$

where ε_a is the relative permittivity of the composite, ε_p is the relative permittivity of the ceramic filler, ε_m is the relative permittivity of the polymer, and V_p is the volume percentage of the filler with respect to the entire composite. This equation is useful in predicting the dielectric properties of a composite in the high-frequency range where a clear interfacial polarization does not appear. The composition of the composite is reflected in k, which is determined on the basis of theoretical equations and experimental data. In a parallel capacitor model, the maximum relative permittivity is obtained at $k = 1$, whereas in a series capacitor model, the minimum relative permittivity is obtained at $k = -1$. Figure 5.4 shows the relationship between the relative permittivity and the volume percentage of the filler calculated using equations of typical models, assuming that the relative permittivities of the polymer and filler are 3 and 9, respectively. The random arrangement model is the geometric mean of the parallel and series capacitor models. The relative permittivity obtained using the random arrangement model takes an intermediate value between the two models for $-1 \leq k \leq 1$; therefore, the equation of the random arrangement model is independent of the type of material used. Furthermore, various empirical equations considering the geometric structure inside the composites have been proposed. Typical empirical equations include the Lichtenecker–Rother and Bruggeman equations. In any model, the relative permittivity of the composite increases with increasing volume percentage of the filler if the relative permittivity of the filler is higher than that of the polymer.

When interfacial polarization develops at the interface between the filler and the polymer, the relative permittivity of the composite can be estimated from the conductivities and relative permittivities of the filler and polymer and their volume percentages. In particular, when a small number of spherical fine particles are dispersed in a medium, the Maxwell–Wagner theory can be applied to predict the dielectric characteristics of the composite in the low-frequency range where a clear interfacial polarization appears.

Which relative permittivity, high or low, is better for electrical insulators? The answer depends on the type of device that uses the insulating material. When used in capacitors, a high relative permittivity is better so that the capacitors can store more charges. However, in coatings and the spacers that support high-voltage conductors, such as motor windings, and a gas-insulating circuit breaker a low relative permittivity is better. Because gas with a low relative permittivity exists around the solid insulating material of the device, electric field concentrates on the gas side owing to the difference in relative permittivity between the gas and the solid, thus inducing discharge. Therefore, a low relative permittivity closest to that of the gas is desired for the solid insulating material.

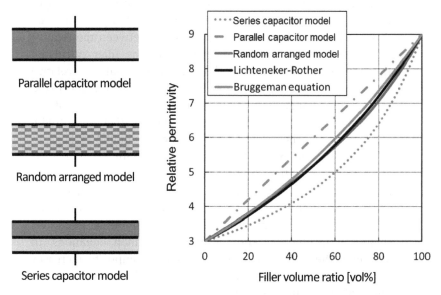

Figure 5.4 Relationship between relative permittivity and volume percentage of the filler calculated using equations of typical models.

5.1.3 Permittivity Rises or Falls Due to Nanofiller Inclusion

When a several-wt% Nanofiller is added to the polymer, the relative permittivity of the composite increases in some cases and decreases in other cases. Figure 5.5 shows the relative permittivity of the nanocomposites obtained by adding 5 wt% nanoclay to different types of epoxy resin [1].

The relative permittivity of the nanocomposite including acid anhydride-cured epoxy resin increases upon adding the nanofiller. In contrast, the relative permittivity of the nanocomposite including amine-cured epoxy resin decreases upon adding the nanofiller. This finding indicates that the effect of using nanocomposites on relative permittivity differs among epoxy resins cured using different curing agents even when the same nanoclay is added. The existence of a huge number of interfaces between nanoclay particles is considered to affect the dielectric property of the epoxy resin. Table 5.2 shows the mechanism behind the effect of nanofiller interface on the dielectric properties of polymers. Note that the effect of using nanocomposites on the dielectric properties differs depending on the type of polymer, such as polyethylene or polypropylene, in addition to epoxy resin.

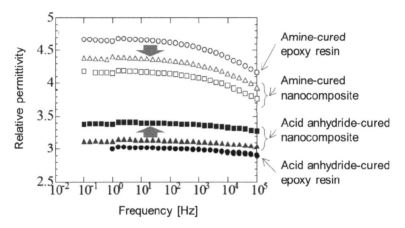

Figure 5.5 Relative permittivity of the nanocomposites obtained by adding 5 wt% nanoclay to different types of epoxy resin [1].

Table 5.2 Mechanism behind the effect of nanofiller interface on the dielectric properties of polymer

Increasing permittivity of polymer (Encouragement of polarization)	Decreasing permittivity of polymer (Suppression of polarization)
Inclusion of impurity ion [2]	Suppression of molecular motion [5]
Breaking of molecular symmetry-structure [3]	Suppression of ion movement [6]
Accumulation of charge on interface defect [4]	Effect of high conductive layer on interface [7]

The relative permittivity of nanocomposites is lower than that of microcomposites containing the same amount of filler.

Figure 5.6 shows the relative permittivities of nanocomposites (obtained by adding a magnesia nanofiller to low-density polyethylene) and microcomposites (obtained by adding a magnesia microfiller to low-density polyethylene). The amount of added filler is 0.5–10 parts per hundred of resin (phr) for both composites [8]. Because the relative permittivity of magnesia is 9.8, which is higher than that of low-density polyethylene (2.2), the relative permittivity of nanocomposites increases with increasing amount of added magnesia nanofiller compared with that of polyethylene. This is because the volume percentage of the filler with a high relative permittivity increases in the composite, as explained in Section 5.2. The same phenomenon is also observed in microcomposites. However, the relative permittivity of nanocomposites is lower than that of microcomposites despite of the same amount of added filler. This is an example of how a large number of interfaces are formed between nanofiller particles affecting the dielectric properties of low-density polyethylene, as indicated previously.

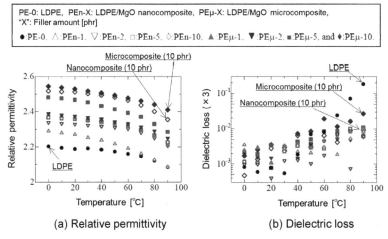

(a) Relative permittivity (b) Dielectric loss

Figure 5.6 Relative permittivities and dielectric loss of nanocomposites (obtained by adding a magnesia nanofiller to low-density polyethylene) and microcomposites (obtained by adding a magnesia microfiller to low-density polyethylene). © 2008 IEEE. Reprinted, with permission, from Ishimoto, K., Tanaka, T., Ohki, Y., Sekiguchi, Y., Murata, Y., and Gosyowaki, M. (2008). Comparison of Dielectric Properties of Low-Density Polyethylene/Mgo Composites with Different Size Fillers, *Annual Rept. IEEE CEIDP*, pp. 208–211.

The dielectric loss of nanocomposites is less than that of low-density polyethylene alone as temperature increases. This is because the motion of the molecular chains of polymers, which are more active at high temperatures, is suppressed by the nanofiller. A similar phenomenon is observed in the composites containing both nanofillers and microfillers [9]. The effects of microfillers and nanofillers on the dielectric loss are complex, and are currently being studied.

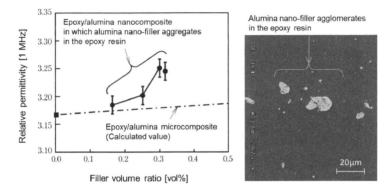

Figure 5.7 Relative permittivity of nanocomposites in which alumina nanofiller agglomerates with a size of several ten μm remain in the epoxy resin. © 2010 IEEE. Reprinted, with permission, from Kurimoto, M., Okubo, H., Kato, K., Hanai, M., Hoshina, Y., Takei, M., and Hayakawa, N. (2010). Dielectric Properties of Epoxy/Alumina Nano-Composite Influenced by Control of Micrometric Agglomerates, *IEEE Trans. Dielectr. Electr. Insul.*, **17**(3), pp. 662–670.

When nanofillers agglomerate or water is absorbed at the interfaces, the relative permittivity of the composites increases. In these cases, water molecules with a high relative permittivity of 80 are trapped by a large number of nanofiller interfaces, leading to an increase in the relative permittivity of nanocomposites depending on the materials included [10]. Techniques for controlling the moisture and for hydrophobic treatment at the surface of nanofillers will be required for some materials, because moisture reduces the withstand voltage. Furthermore, the nanofiller tends to aggregate because it has a large number of interfaces. In particular, agglomerates are formed around moisture as a nucleus. Figure 5.7 shows the relative permittivity of nanocomposites in which alumina nanofiller agglomerates with a size of several ten μm remain in the epoxy resin at 1 MHz [11]. The relative permittivity of the nanocomposites containing the nanofiller agglomerates is higher than that of microcomposites. With decreasing number of nanofiller agglomerates, the relative permittivity of the nanocomposites approaches that of microcomposites.

As explained above, the large number of nanofiller interfaces not only affects the dielectric properties of the polymer but also form insulation defects such as those formed through water absorption and agglomeration, affecting the dielectric properties of the nanocomposites. Thus, the dielectric properties can serve as indices for evaluating the properties of nanocomposites.

5.1.4 Much Attention Is Paid to an Unexpected Strange Phenomenon of Permittivity Decrease

The relative permittivity of nanocomposites obtained by adding a several-wt% nanofiller with a high relative permittivity to the polymer is lower than that of the polymer alone in some cases. Figure 5.8 shows the relative permittivities of nanocomposites obtained by adding a 2 wt% nanofiller (e.g., silica, aluminum nitride, alumina, or magnesia) to epoxy resin [12]. The relative permittivity of the obtained nanocomposites is lower than that of the epoxy resin, although the relative permittivity of the nanofiller is higher than that of the epoxy resin. This phenomenon is unique to nanocomposites and cannot be explained using Eq. (5.2). Similar phenomena are observed for the combinations of epoxy/nanoclay [13], epoxy/titania [5, 14], epoxy/zirconia [14], and polyamide/nanoclay. The possibility of the nanofiller suppressing the molecular movement of the polymer that exhibits dipole polarization is pointed out, because the decrease in relative permittivity is observed in a particularly high-frequency range of several MHz, as shown in Fig. 5.8.

With further increase in the amount of added nanofiller, the relative permittivity, which once decreases, increases again. Figure 5.9 shows the relative permittivity of the nanocomposites at 1 MHz obtained by adding different amounts of alumina nanofiller to epoxy resin [15]. The reason for the decrease in relative permittivity to its minimum value, followed by an increase with increasing amount of added nanofiller, is considered as follows. There are polymer layers with a low relative permittivity where the polarization movement is suppressed by the nanofiller, and these polymer layers overlap each other with increasing amount of added nanofiller. The decrease in relative permittivity uniquely observed in nanocomposites is considered to be caused by the addition of a nanofiller. This effect is useful in determining the amount of nanofiller that should be added; however, this effect is not always observed in combinations of nanofillers and polymers. This is because the effect is affected by the formation pattern of the interfaces, the curing conditions, and the degree of dispersion of the filler.

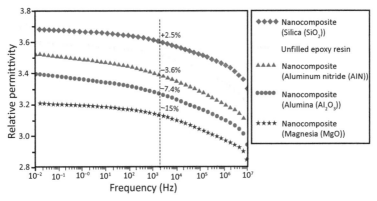

Figure 5.8 Relative permittivity of nanocomposites obtained by adding a 2 wt% nanofiller (e.g., silica, aluminum nitride, alumina, or magnesia) to epoxy resin [12].

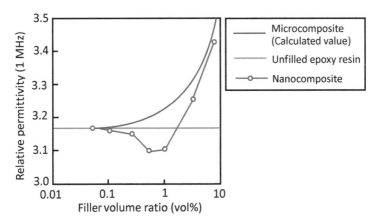

Figure 5.9 Relative permittivity of the nanocomposites at 1 MHz obtained by adding different amounts of alumina nanofiller to epoxy resin [15].

5.1.5 How Do Nanofillers Act for Permittivity of Nanocomposites?

As explained in previous sections, the relative permittivity of the nanocomposite obtained by adding nanofillers to the polymer increases in some cases and decreases in other cases. The mechanisms behind these behaviors that have been clarified thus far are as follows:

(i) The change in relative permittivity caused by the increase in the volume percentage of nanofillers (volume effect).

(ii) The changes in the dielectric properties of polymers as a result of the effect of nanofiller interfaces (interface effect).

The volume effect of the nanofiller mainly follows Eq. (5.2), whereas the mechanisms shown in Table 5.2 are proposed for the interface effect of the nanofiller. Both volume and interface effects have been studied. In particular, the effect of nanofillers of decreasing the relative permittivity of the polymer, as explained in Section 5.4, is unique to nanocomposites. The current understanding of this effect is as follows.

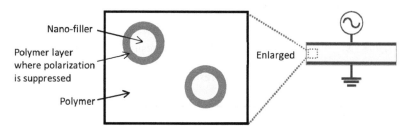

Figure 5.10 Schematic of the mechanism of nanofillers to decrease the relative permittivity of the polymer.

Figure 5.10 shows the schematic of the mechanism of nanofillers to decrease the relative permittivity of the polymer. When nanofillers are added to the polymer, a composite dielectric is formed. In the vicinity of nanofiller interfaces, polymer layers that strongly bind to the nanofiller are formed, and will play a significant role. In these polymer layers in the vicinity of nanofiller interfaces, ion transfer is suppressed, as is the molecular movement of the polymer. Because of this, layers with a low relative permittivity, where polarization is suppressed, appear.

The nanocomposite is a composite dielectric consisting of three materials, i.e., nanofiller, polymer, and polarization-suppressed polymer layer. The relative permittivities of the three materials are as follows: nanofiller > polymer > polarization-suppressed polymer layer. The relative permittivity of the polarization-suppressed polymer layer affects that of the entire composite and the low permittivity unique to nanocomposites is thus observed.

References

1. Tagami, Y., Okada, M., Hirai, N., Tanaka, T., Ohki, Y., Imai, T., Harada, M., and Ochi, M. (2007). Effects of Curing and Filler Dispersion Methods on Dielectric Properties of Epoxy-Nanocomposites, *The Annual Meeting Record IEE Japan*, No. 2–20, p. 21 (in Japanese).

2. Imai, T., Sawa, F., Ozaki, T., Shimizu, T., Kido, R., Kozako M., and Tanaka, T. (2006). Influence of Temperature on Mechanical and Insulation Properties of Epoxy-Layered Silicate Nanocomposite, *IEEE Trans. Dielectr. Electr. Insul.*, **13**(1), pp. 445–452.

3. Tagami, N., Okada, M., Hirai, N., Ohki, Y., Tanaka, T., Imai, T., Harada, M., and Ochi, M. (2008). Dielectric Properties of Epoxy/Clay Nanocomposites -Effects of Curing Agent and Clay Dispersion Method, *IEEE Trans. Dielectr. Electr. Insul.*, **15**(1), pp. 24–32.

4. Mingyan, Z., Tiequan, D., Shujin, Z., Jinghe, Z., Yong, F., Xiaohong Z., and Qingquan, L. (2005). Synthesis and Electric Properties of Nano-Hybrid Polyimide/Silica Film, *Proc. IEEJ ISEIM*, No. P1-27, pp. 397–400.

5. Nelsons, J. K., Fothergill, J. C., Dissado, L. A., and Peasgood, W. (2002). Towards an Understanding of Nanometric Dielectrics, *Annual Rept. IEEE CEIDP*, pp. 295–298.

6. Roy, M., Nelson, J. K., Schadler, L. S., Zou, C., and Fothergill, J. C. (2005). The Influence of Physical and Chemical Linkage on The Properties of Nanocomposites, *Annual Rept. IEEE CEIDP*, pp. 183–186.

7. Fothergill, J. C., Nelson, J. K., and Fu, M. (2004). Dielectric Properties Of Epoxy Nanocomposites Containing TiO_2, Al_2O_3 And ZnO Fillers, *Annual Rept. IEEE CEIDP*, pp. 406–409.

8. Ishimoto, K., Tanaka, T., Ohki, Y., Sekiguchi, Y., Murata, Y., and Gosyowaki, M. (2008). Comparison of Dielectric Properties of Low-Density Polyethylene/Mgo Composites with Different Size Fillers, *Annual Rept. IEEE CEIDP*, pp. 208–211.

9. Iyer, G., Gorur, R. S., Richert, R., Krivda, A., and Schmidt, L. E. (2011). Dielectric Properties of Epoxy Based Nanocomposites for High Voltage Insulation, *IEEE Trans. Dielectr. Electr. Insul.*, **18**(3), pp. 659–666.

10. Zhang, C., and Stevens, G. C. (2008). The Dielectric Response of Polar and Non-Polar Nanodielectrics, *IEEE Trans. Dielectr. Electr. Insul.*, **15**(2), pp. 606–617.

11. Kurimoto, M., Okubo, H., Kato, K., Hanai, M., Hoshina, Y., Takei, M., and Hayakawa, N. (2010). Dielectric Properties of Epoxy/Alumina Nano-Composite Influenced by Control of Micrometric Agglomerates, *IEEE Trans. Dielectr. Electr. Insul.*, **17**(3), pp. 662–670.

12. Kochetov, R., Andritsch, T., Morshuis, P. H. F., and Smit, J. J. (2012). Anomalous Behaviour of The Dielectric Spectroscopy Response of Nanocomposites, *IEEE Trans. Dielectr. Electr. Insul.*, **19**(1), pp. 107–117.

13. Imai, T., Hirano, Y., Hirai, H., Kojima, S., and Shimizu, T. (2002). Preparation and Properties of Epoxy-Organically Modified Layered Silicate Nano-Composites, *Proc. IEEE ISEI*, No. 1, pp. 379–383.

14. Singha, S., and Thomas, M. J. (2008). Dielectric Properties of Epoxy Nano-Composites, *IEEE Trans. Dielectr. Electr. Insul.*, **15**(1), pp. 12–23.

15. Hayakawa, N., Kurimoto, M., Fujii, Y., Shimomura, J., Hanai, M., Hoshina, Y., Takei, M., and Okubo, H. (2010). Influence of Dispersed Nanoparticle Content on Dielectric Property in Epoxy/Alumina Nanocomposites, *Annual Rept. IEEE CEIDP*, pp. 572–575.

5.2 Low Electric Field Conduction

Kazuyuki Tohyama

Department of Electronic Control System Engineering,
National Institute of Technology, Numazu College,
Numazu-shi, Shizuoka 410-8501, Japan

tohyama@numazu-ct.ac.jp

In order to explain the effect of nanocomposite functionalization on the dielectric properties of a polymer, the interfacial phenomenon of the polymer and the filler is considered. Although the expressions of mechanisms in different papers are somewhat different, they have intrinsically the same or different explanations. Some cases of nanocomposite functionalization result in conductivity increase, and others result in conductivity decrease. This behavior is like permittivity, which is mentioned in Section 5.1. This section describes the effects of nanocomposite functionalization and its low electric field conduction.

5.2.1 Electrical Conductivity Is One of the Most Important Factors in Electrical Insulation

When DC voltage is applied to an insulation using parallel-plane electrodes, quite small current flows through it. When estimating the conductivity of insulation, accurate measurement is required, because the resistance of an insulator is extremely high compared with that of the surrounding medium. Thus, the current that flows on the insulation surface should be eliminated from the measurement. Methods for evaluating conductivity ρ are determined by standards such as IEC (International Electrotechnical Commission) and ASTM (American Society for Testing and Materials). When a DC voltage is applied to the insulation sandwiched by two parallel-plane electrodes, most electric lines pass between the electrodes, as shown in Fig. 5.11. In this case, the direction of the electric field between the two electrodes is perpendicular to them. However, as the electric field around the edge of each electrode generates an electric field disturbance, this disturbance induces a radial component of the electric field. As a result, the electric field at the edge of the electrode includes this radial component. Thus, some current flows on the insulation surface, where the resistance is smaller than in the bulk, as shown in Fig. 5.11. This causes an error in the current in the bulk.

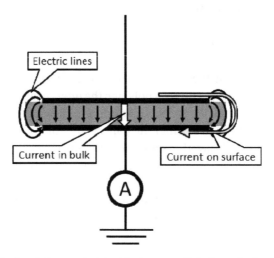

Figure 5.11 Insulator sandwiched by two parallel-plane electrodes.

The evaluation method for volume resistivity ρ in solid insulation is shown in Fig. 5.12. The guard electrode protects the measurement value from the current on the sample sheet surface, so that the volume resistivity ρ in the solid insulation is measured with high accuracy. Figure 5.13 and Table 5.3 show the recommended test piece design for volume resistivity defined by standards. It is not treated in detail here. The volume resistivity is also affected by the surrounding conditions such as temperature and humidity, even though the applied voltage is low. Thus, it is necessary to perform the measurements under defined surrounding temperature and humidity.

Volume resistance $R_V = (V/I)$ is obtained from the applied voltage V and the current I. Furthermore, the volume resistivity ρ_V is obtained from the sample sheet thickness t and Eq. (5.3) as follows:

$$\rho_V = R\left(\pi \frac{D_1^2}{4t}\right) \tag{5.3}$$

The inverse of volume resistivity ρ (Ωm) is the conductivity σ (S/m) as shown in Eq. (5.4).

$$\sigma = \frac{1}{\rho_v} \tag{5.4}$$

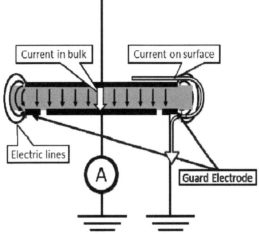

Figure 5.12 Evaluation of volume resistivity of insulator using three-terminal electrode system.

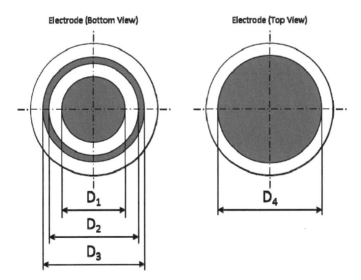

Figure 5.13 Recommended test piece design of volume resistivity.

Table 5.3 Test piece and electrode size

	Recommended size		Bottom (mm)			Top (mm)
Standard	Diameter D	Thickness t	D_1	D_2	D_3	D_4
ASTM D257	100	3	76	88	100	100
	50	3	25	38	50	50
JIS K 6911	100	2	50 ± 0.5	70 ± 0.5	80 ± 0.5	83 ± 2
IEC 60093	100	2.5	50 ± 0.5	60 ± 0.5	80 ± 0.5	83 ± 2

5.2.2 Electrical Conductivity Increased by Nanofiller Addition in Some Cases

Increase in electrical conductivity due to nanofiller addition into a polymer is undesirable. However, some cases have been reported and their mechanisms are proposed.

For example, conductivity, σ, of polyimide/alumina (PI/Al$_2$O$_3$) nanocomposite, shown in Fig. 5.14, increases exponentially with nanofiller content. A PI/Al$_2$O$_3$ nanocomposite film is made by a material synthesizing method called "sol-gel process," in which polyimide and alumina nanofiller are synthesized. In the figure, the vertical axis indicates conductivity σ on a logarithmic scale, and it is found that only 10% nanofiller addition creates a conductivity increment of one digit.

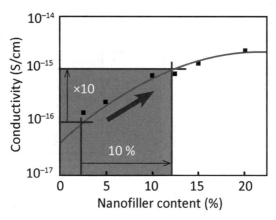

Figure 5.14 Effect of Al$_2$O$_3$ content on direct current conductivity of PI/Al$_2$O$_3$ nanocomposite polyimide films [1].

Furthermore, though its electric field is slightly higher than the previous case, for nanoclay addition into EVA (ethylene vinyl-acetate) and iPP (isotactic polypro-pylene), conductivity increases have been reported. Conductivity σ is the product of charge density n and mobility μ. As shown in Fig. 5.15, the time variations of mobility μ for EVA and iPP clearly indicate mobility increase resulting from nanofiller addition: about 1 and 2–3 digits for EVA and iPP, respectively.

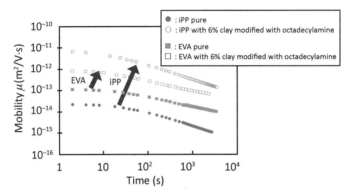

Figure 5.15 Estimated apparent trap-controlled mobility μ taken from depolarization characteristics for iPP and EVA, poling field 60 kV/mm [2].

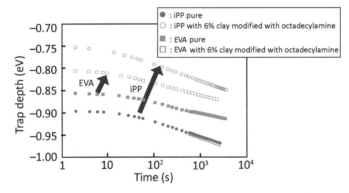

Figure 5.16 Estimated trap depth distribution in EVA and iPP taken from apparent trap-controlled mobility characteristics of Fig. 5.15 [2].

Figure 5.16 shows the estimated trap depth distribution in EVA and iPP taken from the apparent trap-controlled mobility characteristics of Fig. 5.15. Here, trap depth is a physical property relating to charge density n. It means that the trap depth is shallower, and increased charge density n clearly contributes to conduction. Here, it is explained that nanofiller addition introduces shallow traps in the polymer.

5.2.3 Electrical Conductivity Decreases with Nanofiller Addition in Some Cases

Nanofiller addition into insulating materials is often used to improve insulation performance by decreasing electrical conductivity, elevating electrical breakdown strength, improving partial discharge resistance, and so on. The materials listed in Table 5.4 are polymer/nanocomposite dielectric materials, which are reported to decrease electrical conductivity by nanofiller addition.

Proposed mechanisms of this decrease are shown in Table 5.4. For example, it is explained that carriers that are captured at the trap level within a forbidden ban, based on the "Band Theorem," cause a decrease in charge density, which contributes to conduction. The proposed mechanism of drastic variation of conductivity is explained by the formation of an interaction zone caused by nanocomposite functionalization, and the percolation effect. As shown in Fig. 5.17, reducing the filler size from micro to nano, or chemically treating the filler surface, will bring about a significant change in conductivity. This means that the physical properties of the interface between the nanofiller and the polymer strongly influence the physical properties of the nanocomposite materials themselves. The "double layer model" and "multi-core interfacial model" are proposed and discussed. Why is the physical property of "interface" so important? This will be explained from the viewpoint of electrical conduction.

To make the story simple, it is explained by the "double layer model." Here, a spherical microparticle as shown in Fig. 5.18 will be considered. In this case, an interaction zone exists at the interface between the microparticles and the polymer matrix. If this spherical microparticle size is reduced to spherical nanoparticle size of the same quantity, though each particle size becomes smaller by a few digits, the number of spherical nanoparticles will increase by a few digits. As a result, the distances between nanoparticles become shorter than those between microparticles. As shown in Fig. 5.18, although a pass within the interaction zone will be formed, the length will become larger than a straight line between two points. In this model, the nanoparticles disturb the charge motion, thus decreasing the electrical conductivity.

Table 5.4 Nanocomposites that make conductivity decrease by nanofiller addition and its mechanism [21]

Nanocomposite	Mechanism	Ref.
PA/Clay	Decrease of conductivity: barrier effect due to suppression of molecular motion	[3–4]
PI/SiO$_2$	Temperature rise in TSC: introduction of deep traps causes conductivity decrease	[5–6]
	Decrease of conductivity: collapse of crystallinity	[2, 7]
XLPE/SiO$_2$	Existence of high conductivity layer near filler surface; conductivity through double-layer and/or hydration effects is altered by enhanced coupling associated with functionalized materials	[8]
XLPE/MgO	Impurity responsible for space charge formation is adsorbed in MgO nanofiller	[9–10]
LDPE/MgO	Mechanisms of conductivity decrease are explained as follows: (1) MgO nanofiller creates a deep trap, causing homo-space charge accumulation near electrode interface, thus relaxing the electric field (2) Molecular motion and hopping are suppressed; new TSC (Thermally Stimulated Current) indicates introduction of traps by nanocomposite functionalization	[11–16]
LDPE/ZnO	Decrease of conductivity is very small with microfiller addition; the effect on electron injection due to nanofiller addition is not confirmed; it should be discussed based on the trapping process to explain the conductivity suppression mechanism	[17]
LDPE/SiO$_2$	Traps due to strong interaction will be produced by nanofiller addition; at the same time, more mobile ions will occur	[18, 20]

PA: polyamide; PI: polyimide; XLPE: cross-linked polyethylene; LDPE: low-density polyethylene; TSC: thermal stimulated current.

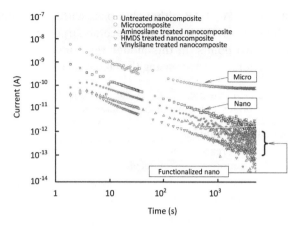

Figure 5.17 Time-dependent conduction for XLPE–SiO$_2$ composites. © 2005 IEEE. Reprinted, with permission, from Roy, M., Nelson, J. K., Schadler, L. S., Zou, C., and Fothergill, J. C. (2005). The Influence of Physical and Chemical Linkage on the Properties of Nanocomposites, *Annual Rept. IEEE CEIDP*, pp. 183–186.

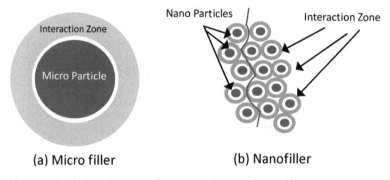

Figure 5.18 Interaction zone between micro- and nanofiller.

The percolation effect was proposed long before the appearance of nanocomposite materials. For example, when an insulating filler and a conductive filler are uniformly mixed, if the insulating filler is rich, the composite material will be insulating, and if the conductive filler is rich, it will be conducting. This variation of volume resistivity is not proportional to the concentration of fillers once it exceeds a certain value. It rapidly varies from insulator to conductor, and vice versa.

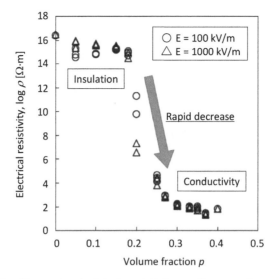

Figure 5.19 Change in electrical resistivity ρ of composites made from Fe_3O_4 and PE as a function of volume fraction p of Fe_3O_4 [22].

Change in electrical resistivity ρ of composites made from Fe_3O_4 and PE as a function of volume fraction p of Fe_3O_4 is shown in Fig. 5.19. According to the theory of the percolation effect, volume resistivity ρ is expressed by Eq. (5.5) as follows:

$$\rho \propto (p - p_c)^{-t} \tag{5.5}$$

Here, p, p_c, and t are the volume fraction, the threshold level of volume fraction, and the critical index, respectively.

References

1. Peihong, Z., Gang, L., Lingyun, G., and Qingquan, L. (2006). Conduction Current Characteristics of Nano-Inorganic Composite Polyimide Films, *Proc. 8th IEEE ICPADM*, pp.755–758.
2. Montanari, G. C., Fabiani, D., and Palmieri, F. (2004). Modification of Electrical Properties and Performance of EVA and PP Insulation through Nanostructure by Organophilic Silicates, *IEEE Trans. Dielectr. Electr. Insul.*, **11**(5), pp. 754–762.
3. Fuse, N., Ohki, Y., Kozako, M., and Tanaka, T. (2008). Possible Mechanisms of Superior Resistance of Polyamide Nanocomposites

to Partial Discharges and Plasmas, *IEEE Trans. Dielectr. Electr. Insul.,* **15**(1), pp. 161–169.

4. Fuse, N., Kikuma, T., Kozako, M., Tanaka, T., and Ohki, Y. (2005). Dielectric Properties of Polymer Nanocomposites, *IEEJ the Paper of Technical Meeting on Dielectrics and Electrical Insulation,* No. DEI-05-59, pp. 9–16 (in Japanese).

5. Cao, Y., Irwin, P. C., and Younsi, K. (2004). The Future of Nanodielectrics in the Electrical Power Industry, *IEEE Trans. Dielectr. Electr. Insul.,* **11**(5), pp. 797–807.

6. Cao, Y., and Irwin, P. C. (2003). The Electrical Conduction in Polyamide Nanocomposite, *Annual Rept. IEEE CEIDP,* pp. 116–119.

7. Mingyan, Z., Tiequan, D., Shujin, Z., Jinghe, Z., Yong, F., Xiaohong, Z., and Qingquan, L. (2005). Synthesis and Electric Properties of Nano-hybrid Polyimide/silica Film *Proc. IEEE ISEIM,* pp. 397–400.

8. Roy, M., Nelson, J. K., Schadler, L. S., Zou, C., and Fothergill, J. C. (2005). The Influence of Physical and Chemical Linkage on the Properties of Nanocomposites, *Annual Rept. IEEE CEIDP,* pp. 183–186.

9. Murata, Y., and Kanaoka, M. (2006). Development History of HVDC Extruded Cable with Nano Composite Material, *Proc. 8th IEEE ICPADM,* pp. 460–463.

10. Maekawa, Y., Yamaguchi, A., Hara, M., and Sekii, Y. (1992). Research and Development of XLPE Insulated DC Cable, *IEEJ Trans. Power and Energy,* **112**(10), pp. 905–913 (in Japanese).

11. Murata, Y., Sekiguchi, Y., Inoue, Y., and Kanaoka, M. (2005). Investigation of Electrical Phenomena of Inorganic-Filler/LDPE Nanocomposite Material, *Proc. IEEJ ISEIM,* **3**, pp. 650–653.

12. Ishimoto, K., Kikuma, T., Tanaka, T., Sekiguchi, Y., Murata, Y., Ohki, Y. (2007). Effect of the Sheet Formation Method on the Electric Conduction Characteristics in Low-Density Polyethylene/MgO Nanocomposites, *IEEJ the Paper of Technical Meeting on Dielectrics and Electrical Insulation,* No. DEI-07-55, pp. 1–6 (in Japanese).

13. Ishimoto, K., Tanaka, T., Ohki, Y., Sekiguchi, Y., Murata, Y., and Gosyowaki, M. (2008). Dielectric Properties of LDPE with MgO Fillers Different in Diameter, *IEEJ the Paper of Technical Meeting on Dielectrics and Electrical Insulation,* No. DEI-08-65, pp. 1–5 (in Japanese).

14. Murakami, Y., Nemoto, M., Kurnianto, R., Hozumi, N., Nagao, M., and Murata, Y. (2006). Space Charge Characteristic of Nano-Composite Film of MgO/LDPE under DC Electric Field, *IEEJ Trans. Fundamentals Mater.,* **126**(11), pp. 1078–1083.

15. Kikuma, T., Fuse, N., Tanaka, T., Murata, Y., and Ohki, Y. (2006). Filler-Content Dependence of Dielectric Properties of Low-Density Polyethylene/MgO Nanocomposites, *IEEJ Trans. Fundamentals Mater.*, **126**(11), pp. 1072–1077.

16. Hinata, K., Fujita, A., Tohyama, K., and Murata, Y. (2006). Dielectric Properties of LDPE/MgO Nanocomposite Material under AC High Field, *Annual Rept. IEEE CEIDP*, pp. 313–316.

17. Fleming, R. J., Ammala, A., Casey, P. S., and Lang, S. B. (2005). Conductivity and Space Charge in LDPE/BaSrTiO$_3$ Nanocomposite, *IEEE Trans. Dielectr. Electr. Insul.*, **12**(4), pp. 745–753.

18. Chen, J., Yin, Y., Li Z., and Xiao, D. (2005). Study the Percolation Phenomenon of High Field Volt-Ampere Characteristic in the Composite of Low-density Polyethylene/Nano-SiO$_x$, *Proc. IEEJ ISEIM*, pp. 243–246.

19. Yin, Y., Chen, J., Li, Z., and Xiao, D. (2005). High Field Conduction of the Composite of Low-Density Polyethylene/nano SiO$_x$ and Low-Density Polyethylene/Micrometer SiO$_2$, *Proc. IEEJ ISEIM*, pp. 405–408.

20. Yin, Y., Dong, X., Chen, J., Li, Z., and Dang, Z. (2006). High Field Electrical Conduction in the Nanocomposite of Low-Density Polyethylene and Nano-SiO$_x$, *IEEJ Trans. Fundamentals Mater.*, **126**(11), pp. 1064–1071.

21. Investigating R&D Committee on Polymer Nanocomposites and Their Applications as Dielectrics and Electrical Insulation (2009). Characteristics Evaluation and Potential Applications of Polymer Nanocomposites as Evolutional Electrical Insulating Materials, *IEEJ Technical Report*, No. 1148, pp. 19–21.

22. Okamoto, T., Kawahara, M., Yamada, T., Inoue, Y., and Nakamura, S. (2006). Percolation Phenomena of the Composites Made with Two Kinds of Filler and a New Potential Grading Materials, *IEEJ Trans. Fundamentals Mater.*, **126**(10), pp. 1004–1012 (in Japanese).

5.3 Conduction Current under High Electric Field and Space Charge Accumulation

Yasuhiro Tanaka

Department of Mechanical Systems Engineering,
Tokyo City University, Setagaya-ku, Tokyo 158-8557, Japan

ytanaka@tcu.ac.jp

Under high electrical stress, conduction current in an insulating material increases, sometimes causing space charges to accumulate in its bulk. Accumulated space charges distort electric field distribution, which sometimes induces degradation of the material leading to a fatal electric breakdown. To resolve this problem, nanocomposite technology may be applied to improve insulating materials.

5.3.1 Is Breakdown Unpredictable?

When a high DC voltage is applied to a solid insulating material, a very small conduction current flows in it. When this current is measured using a circuit shown in Fig. 5.20, a relatively large "momentary" I_0 [A] $(=V_0/R)$ is measured at first as shown in Fig. 5.21. After that, it gradually decreases with time and finally vanishes if the material is assumed as an ideal capacitor. However, in the case of actual insulating materials, the current does not vanish, but finally becomes a small steady current I_d after a very long time. This steady current is called "DC leakage current" [1].

Current density J [A/m^2] (current per unit area) of this DC leakage current is proportional to an electric field E [V/m] (applied voltage per thickness of the material) when the applied voltage is relatively low as shown in Fig. 5.22. The proportional relation is called "Ohm's law" and the proportional coefficient is called "conductivity" σ [S/m]. In other words, the relationship is described as $J = \sigma E$. For insulating materials, this range of electric field where Ohm's law is applicable is defined as "low electric field" [1].

Conduction Current under High Electric Field and Space Charge Accumulation | 211

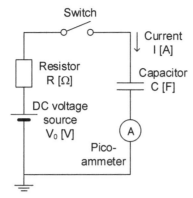

Figure 5.20 Measurement circuit for conduction current in dielectric materials.

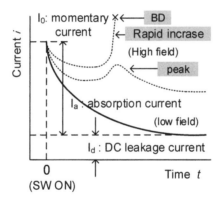

Figure 5.21 Time dependences of conduction current.

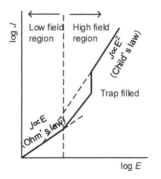

Figure 5.22 Relationship between conduction current (J) and applied electric field (E).

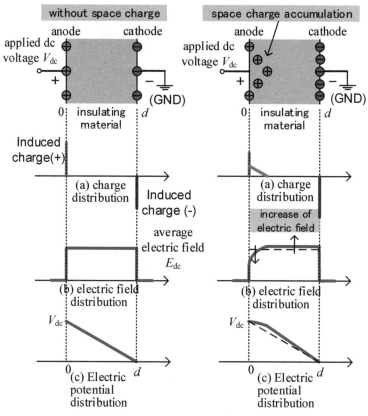

Figure 5.23 Schematic models of (a) space charge density, (b) electric field and (c) electric potential distributions.

However, when the applied voltage is increased, the current density increases, thus deviating from Ohm's law. This range of electric field where the current density deviates from Ohm's law is generally recognized as a "high electric field." The threshold between the ranges of low and high electric fields depends on the materials. One reason for the deviation from Ohm's law under a high electric field is thought to be that a so-called "space charge" accumulates in the material. Figures 5.23a–c show schematic models of the charge (top), electric field (middle) and electric potential (bottom) distributions under applied stresses of low and high electric fields, respectively. As shown in Fig. 5.23a, when the voltage is applied to the insulating material, positive and negative charges are induced on anode and cathode electrodes by the applied voltage. Consequently, the electric field in the bulk of the material is homogeneous through the bulk, as shown in Fig. 5.23a. On the other hand, as shown in Fig. 5.23b, when the space charge accumulates in the bulk of the material, the electric field is distorted by it and an area where the electric field is larger or smaller than the average electric field is generated in the material. In the material, it is said that the charges are "trapped" in so-called "trap sites." With an increase in the amount of trapped charge, distortion of the electric field is increased somewhere, and then the conduction current consequently increases, deviating from Ohm's law.

It is generally said, however, that since it takes a certain time for the accumulation of the space charge, it does not affect the conduction current under AC stress or short duration of voltage application. Furthermore, as a matter of fact, it is hard to measure the stable DC leakage current under high electric field because the time-dependent current sometimes shows a peak or a rapid increase followed by an electrical breakdown. Such phenomena are unpredictable, which makes it difficult to recognize the characteristics of the materials under high electric field.

5.3.2 Is Space Charge Accumulation Presage of Degradation or an Electrical Breakdown?

While the relationship between space charge accumulation and breakdown under high DC electric stress was unclear for a long time, it has become increasingly obvious because some advanced systems for measuring space charge distribution have been developed in recent years.

Figure 5.24 shows the observed space charge accumulation process followed by an electrical breakdown in LDPE (low-density polyethylene) film (150 μm-thick) under an applied voltage of 50 kV, which corresponds to an average electric field of 330 kV/mm (=3.3 MV/cm). In Fig. 5.24a, the horizontal and vertical axes indicate the time of voltage application and the position in the thickness direction of the film, respectively [2]. The color distribution in the figure indicates the charge density. (The red and blue colors indicate positive and negative charges, respectively.) As shown in the figure, it is found that a large amount of positive charge appeared near the anode (right-hand side in the figure) just after the voltage application, and it moved toward the cathode (left-hand side), keeping a packet-like shape. However, the movement of the charge gradually slowed down, and stopped at the middle of the bulk, then finally an electrical breakdown was observed 20 min after the voltage application. A Japanese researcher discovered the existence of the large packet-shaped charge using a so-called PEA (pulsed electro-acoustic) method, which was developed in Japan in the 1990s, and called it "packet-like charge" or "charge packet." Figure 5.24c shows the time dependence of an electric field distribution obtained by integral calculation of the measured space charge distributions. Judging from the result, the electric field in the bulk of LDPE near the cathode increased gradually with movement of the packet-like charge. Breakdown was observed when the electric field was about 550 kV/mm, which is much higher than the average field of 330 kV/mm. This result indicates that the breakdown occurred in LDPE under high DC stress after the enhanced zone of the electric field appeared in the bulk.

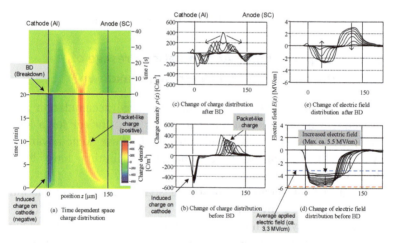

Figure 5.24 Packet-like charge accumulation behavior in LDPE under high DC stress [2].

It is thus found that the large packet-like charge injection increases the electric field in the bulk of LDPE partly under high DC stress. Figure 5.25 shows the relationship between the applied average DC stress and the maximum electric field enhanced by the space charge accumulation [3]. As shown, the maximum electric field in the bulk of the LDPE was close to the applied average DC stress up to about 100 kV/mm. However, when the applied average DC stress was larger than 100 kV/mm, the packet-like charge injection became obvious and the maximum electric field was greatly increased. The result shows the enhanced electric field sometimes doubled the value of the applied stress. It is said that an impulse breakdown strength of LDPE is in the range of 500–600 kV/mm. As mentioned above, since the maximum electric field under a DC stress of 300 kV/mm exceeds 500 kV/mm, the possibility of an electric breakdown in the material increases.

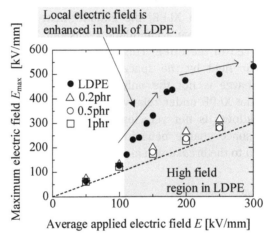

Figure 5.25 Maximum electric field induced by accumulation of space charge in LDPE and LDPE/MgO nanocomposite under various DC stresses [5].

As mentioned above, when the excess DC stress is applied to LDPE, it causes a large amount of space charge accumulation, and the resulting enhanced electric field seems to trigger the breakdown. On the other hand, it has been reported that the breakdown seems to be caused by the space charge accumulation in XLPE (cross-linked polyethylene), which is usually used as an insulating layer of actual AC power transmission cable [4]. Figures 5.26a–d show time dependences of space charge distributions in XLPE under DC stresses of 50, 100, 150, and 200 kV/mm [5]. As shown in Fig. 5.26b, the large amount of positive packet-like charge injection, which is similar to that observed in LDPE, was observed under a DC stress of 100 kV/mm.

However, as shown in Fig. 5.26c, the space charge accumulation behavior was completely different from that observed in LDPE. In the case of XLPE under a DC stress of 150 kV/mm, while the injected positive charge also moved towards the cathode side as observed in LDPE, it finally reached the cathode. After that, it disappeared at the cathode and then reappeared from the anode. This process was repeatedly observed in this measurement. While such a repeated charge injection was also observed in LDPE for a measurement of several hours, it usually occurred for a relatively short duration of a few minutes in XLPE. While the electric breakdown sometimes followed the repetition of the charge injection in XLPE, the enhancement of the electric field by the space charge accumulation was not large compared to that observed in the LDPE. This means that the enhancement of the electric field by the space charge accumulation of the packet-like charge is not the only reason why the breakdown occurred in the XLPE under relatively low DC stress. The reason for the breakdown is not yet clear. However, it is natural that the curious space charge behavior observed in XLPE must be closely related to the breakdown mechanism in XLPE.

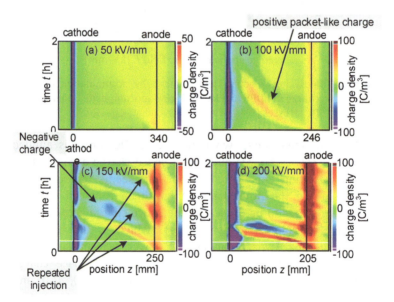

Figure 5.26 Space charge accumulation behavior in XLPE under various applied DC stresses [8].

As mentioned above, under the high DC stress that induces a large amount of space charge accumulation, breakdown occasionally occurs at a lower strength than the intrinsic breakdown stress that seems to be close to the impulse breakdown strength. This fact annoyed us for a long time when designing high-voltage devices because the space charge had not yet been clearly observed.

5.3.3 Adding Nano-Size Filler to Significantly Reduce Space Charge Accumulation Even under High DC Stress!

As mentioned above, when a very high DC stress is applied to polyethylene, a large amount of a packet-like charge is injected into the bulk of the material. This leads to fatal breakdown due to enhancement of the local electric field or other obscure reasons even under a lower DC stress than the intrinsic breakdown strength. This means that if there is a material in which no packet-like charge accumulates in the bulk even under such a very high DC stress, it may keep up a state as a good insulating material even under a very high DC stress of less than its intrinsic breakdown strength. However, it has become increasingly clear that such a packet-like charge does not accumulate in some kinds of composite materials, including certain kinds of inorganic nano-size fillers.

Figures 5.27a,b show time dependencies of space charge distributions in LDPE and LDPE/MgO, respectively, under an average DC stress of 200 kV/mm at room temperature (ca. 25°C) [3]. The thicknesses of both are about 70 µm. The LDPE/MgO nanocomposite includes MgO nanofiller with a content of 1 phr, which means a rate of 1 g filler content per 100 g base polymer. As shown in Fig. 5.27b, there is no charge accumulation in the bulk of LDPE/MgO, while a large amount of packet-like charge is injected into the bulk of LDPE under the same condition. Therefore, the electric field in LDPE/MgO is not distorted much even under such a high DC stress. Figure 5.25 also shows the relationship between the applied average DC stress and the maximum electric field in LDPE/MgO enhanced by the space charge accumulation. As shown in this figure, the maximum electric field in LDPE/MgO is close to the applied average electric field even under the very high stress of 300 kV/mm.

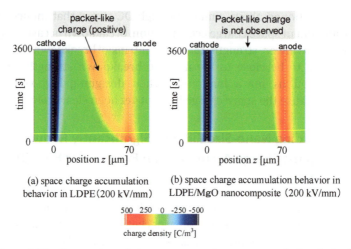

Figure 5.27 Comparison between space charge accumulation behaviors in (a) LDPE and that in (b) LDPE/MgO under high DC stress [5].

It has also been reported that the packet-like charge injection is effectively suppressed by adding nano-size filler to XLPE. Figures 5.28a,b show time dependencies of space charge distributions in XLPE and SXL-A, respectively, under an average DC stress of 200 kV/mm at room temperature (ca. 25°C) [5]. SXL-A is a nanocomposite material including an inorganic conductive nano-size filler. As shown in Fig. 5.28b, no obvious packet-like charge was observed in SXL-A, while the repeated charge packet-like charge was injected in XLPE under the same condition. The electric field in SXL-A is not distorted much even under such high DC stress. Figure 5.29 shows the relationship between the applied average DC stress and the maximum electric field enhanced by the space charge accumulation in XLPE and SXL-A. In this figure, the sample with a filler content of 0.0 wt% indicates the maximum electric field in XLPE. While it is found that electric field enhancement is observed in XLPE under an applied stress of more than 100 kV/mm, it is not obviously observed in SXL-A. Therefore, it is expected that breakdown due to the accumulation of the packet-like charge will not occur in such a material.

It is remarkable that a relatively small amount of filler such as 1 phr (Fig. 5.25) or 0.1 wt% (Fig. 5.29) added to a base material dramatically improves the space charge accumulation property of the material. For insulating materials of power cable, mechanical properties like flexibility are also important. However, since the addition of such a small amount of filler does not affect the mechanical properties at all, the dramatic improvement of insulating properties with the addition of nano-size filler is extremely effective. In fact, a newly developed DC power transmission cable (±250 kV, 45 km) using XLPE improved by addition of inorganic nano-size filler was installed between Honshu (main island) and Hokkaido (northern island) of Japan in December 2012, and it has since been operating safely and stably [5].

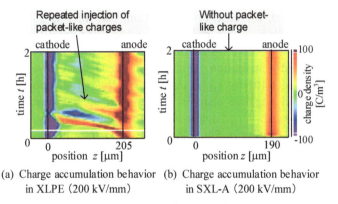

(a) Charge accumulation behavior in XLPE (200 kV/mm)   (b) Charge accumulation behavior in SXL-A (200 kV/mm)

Figure 5.28 Comparison between space charge accumulation behaviors in (a) XLPE and that in (b) SXL-A under high DC stress [8].

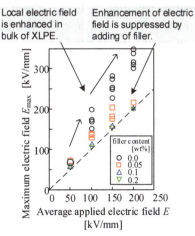

Figure 5.29 Maximum enhanced electric field by charge accumulation in XLPE and SXL-A under various applied electric fields [8].

5.3.4 Why Does Addition of Nano-Size Filler Suppress Injection of Packet-Like Charge?

Why does the addition of inorganic nano-size filler suppress the injection of the packet-like charge into the bulk of a nanocomposite material? In fact, details of the suppression mechanism remain to be completely elucidated, but some mechanisms have been proposed. The following describes one of the current postulated mechanisms.

Figure 5.30 shows schematic models of Faraday's line of electric force around a filler added to a polymer under DC stress. In this model, for ease of explanation, the filler is assumed as spherical with a certain radius. When the filler permittivity ε_{r2} is equal to that of the base polymer ε_{r1}, there is no distortion of the electric field distribution, as shown in Fig. 5.30a. However, when the filler permittivity ε_{r2} is larger than that of the polymer ε_{r1}, positive and negative charges are induced on the surface of the filler, and then a large distortion of the field is generated around the filler as shown in Fig. 5.30b. Since the permittivities of MgO and LDPE are about 9.8 and 2.3, respectively, this applies to the LDPE/MgO nanocomposite. If the filler permittivity ε_{r2} is much larger than that of the polymer ε_{r1}, for example if the filler is conductive, the distortion of the field is tremendous.

When electric field distortion generates around the filler, the potential distribution around it is also distorted. Figure 5.31 shows its distorted distribution. It is found that "potential wells," which can play roles of carrier traps for positive and negative charges under a DC stress, are located around the filler as shown in the figure. When a carrier drifts beside the filler, the carrier, whether it is positive or negative, must be captured by the well, and it cannot move anymore. The depth of the well increases when the difference between the permittivities increases or the applied stress is larger. When the depth increases, its capture cross section increases. When a DC stress is applied to such a material and the charge is injected into its bulk from the electrode, the injected charge is captured by the deep trap near the injected surface and it cannot move from there. Therefore, a large amount of charge is accumulated at the surface, which reduces the electric field at the electrode/material interface. The reduction of the electric field must induce a suppression of the charge injection.

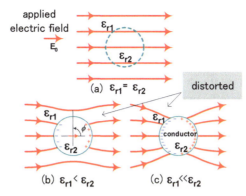

Figure 5.30 Schematic models of line of electric force around filler added in polymers under DC stress. © 2008 IEEE. Reprinted, with permission, from Takada, T., Hayase, Y., Tanaka, Y., and Okamoto, T. (2008). Space Charge Trapping in Electrical Potential Well Caused by Permanent and Induced Dipoles for LDPE/MgO Nanocomposite, *IEEE Trans. Dielectr. Electr. Insul.*, **15**(1), pp. 152–160.

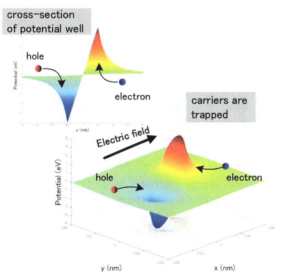

Figure 5.31 Schematic model of electric potential well around filler under DC stress. © 2008 IEEE. Reprinted, with permission, from Takada, T., Hayase, Y., Tanaka, Y., and Okamoto, T. (2008). Space Charge Trapping in Electrical Potential Well Caused by Permanent and Induced Dipoles for LDPE/MgO Nanocomposite, *IEEE Trans. Dielectr. Electr. Insul.*, **15**(1), pp. 152–160.

In this case, since the capture cross section increases when the difference between the permittivities is larger, a relatively small amount of added filler is effective if its permittivity is much larger than that of the base polymer. That may be why even the addition of a relatively small amount of conductive inorganic filler of 0.1 wt% is effective for the suppression of packet-like charge injection into the sample of SXL-A. Furthermore, if this model is correct, when a larger DC stress is applied to the sample, the suppression effect must be greater. Figure 5.32 shows that the space charge distribution in LDPE/MgO (1 phr) with a thickness of about 70 μm under various DC stresses between 50 and 250 kV/mm. The result shows that the injected space charge is distributed shallower (closer to the injected electrode) under a higher stress.

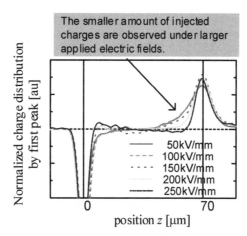

Figure 5.32 Space charge distribution in LDPE/MgO nanocomposite under various applied stresses [10].

In this model, the filler is expected to play a role of a trap site for both positive and negative carriers because negative and positive potential wells are formed around it under a high DC stress, as shown in Fig. 5.31. This means that the trap site is effective not only for carriers injected from the electrode but also for ionic carriers generated from the bulk.

As mentioned above, nanocomposite technology is expected to improve insulating materials under the DC stress, and many new materials will be developed using this method.

References

1. Inuishi, Y., et al. (1973). *Yudentai Gensho-Ron* (Chapter 4 and 5), IEEJ, pp. 203–321 (in Japanese).

2. Matsui, K., Tanaka, Y., Takada, T., Fukao, T., Fukunaga, K., Maeno, T., and Alison, J. M. (2005). Space Charge Behavior in Low-Density Polyethylene at Pre-Breakdown, *IEEE Trans. Dielectr. Electr. Insul.*, **12**(3), pp. 406–415.

3. Li, Y., Yasuda, M., and Takada, T. (1994). Pulsed Electroacoustic Method for Measurement of Charge Accumulation in Solid Dielectrics, *IEEE Trans. Dielectr. Electr. Insul.*, **1**(2), pp. 188–198.

4. Hozumi, N., Takeda, T., Suzuki, H., and Okamoto, T. (1998). Space Charge Behavior in XLPE Cable Insulation under 0.2–1.2 MV/cm DC Fields, *IEEE Trans. Dielectr. Electr. Insul.*, **5**(1), pp. 82–90.

5. Hayase, Y., Aoyama. H., Matsui, K., Tanaka, Y., Takada, T., and Murata, Y. (2006). Space Charge Formation in LDPE/MgO Nano-Composite Film under Ultra-high DC Electric Stress, *IEEJ Trans. Fundamentals Mater.*, **126**(11), pp. 1084–1089.

6. Murata, Y., Sekiguchi, Y., Inoue, Y., Kanaoka and M. (2005). Investigation of Electrical Phenomena of Inorganic filler/LDPE Nano-Composite Material, *Proc. IEEJ ISEIM*, pp. 650–653.

7. Matsui, K., Miyawaki, A., Tanaka, Y., Takada, T., and Maeno, T. (2005). Influence of Space Charge Formation in LDPE and XLPE on Electrode Materials under Near Electric Breakdown Field, *IEEJ the Paper of Technical Meeting on Dielectrics and Electrical Insulation*, No. DEI-05-64, pp. 37–42 (in Japanese).

8. Harada, H., Hayashi, N., Tanaka, Y., Maeno, T., Mizuno, T., and Takahashi, T. (2011), Effect of Conductive Inorganic Fillers on Space Charge Accumulation Characteristics in Cross-Linked Polyethylene, *IEEJ Trans. Fundamentals Mater.*, **131**(9), pp. 804–810 (in Japanese).

9. Takada, T., Hayase, Y., Tanaka, Y., and Okamoto, T. (2008). Space Charge Trapping in Electrical Potential Well Caused by Permanent and Induced Dipoles for LDPE/MgO Nanocomposite, *IEEE Trans. Dielectr. Electr. Insul.*, **15**(1), pp. 152–160.

10. Hayase, Y., Takada, T., Tanaka, Y., and Okamoto, T. (2007). Potential Distribution and Space Charge Suppression Effect by Induced Dipole Polarization of Inorganic Nano Filler, *IEEJ the Paper of Technical Meeting on Dielectrics and Electrical Insulation*, No. DEI-07-56, pp. 7–12 (in Japanese).

5.4 Short-Term Breakdown Characteristics

Yoshinobu Mizutani

Central Research Institute of Electric Power Industry (CRIEPI), Yokosuka-shi, Kanagawa 240-0196, Japan

mizutani@criepi.denken.or.jp

Short-term breakdown characteristics are important in designing insulators with various performance characteristics, such as the withstand voltage (measured in insulation tests on devices using insulators) and robustness against voltage surges caused by lightning or switching events. These characteristics vary when an insulator in the form of a nanocomposite is used. If short-term breakdown characteristics deteriorate, the improved characteristics resulting from the use of the nanocomposite insulator cannot be effectively utilized. Therefore, the factors leading to changes in short-term breakdown characteristics have been discussed.

5.4.1 Method of Measuring Short-Term Breakdown Characteristics

When the intensity of an electric field applied to an insulator exceeds a certain value, the current passing through the insulator abruptly increases to nearly infinity, resulting in electrical breakdown. For solids, heat is generated upon electrical breakdown, causing the materials to irreversibly burn out. Thus, electrical breakdown can be a factor that determines the lifetime of insulators. Therefore, breakdown characteristics are one of the important characteristics of insulators, and the clarification of these characteristics and the breakdown phenomenon has been a target of research. Electrical breakdown that occurs in a short period and its characteristics are important in designing insulators with various performance characteristics, such as the withstand voltage (measured in insulation tests on devices using insulators) and robustness against voltage surges caused by lightning or switching events.

Knowledge of the short-term breakdown characteristics is essential when using nanocomposites as insulators. There have been studies on the improvement of short-term breakdown characteristics using nanocomposite insulators. Research on the fabrication of nano/microcomposite insulators by adding nanoparticles has also been carried out to prevent the deterioration of short-term breakdown characteristics when microcomposite insulators are used to improve other characteristics.

Short-term breakdown characteristics are determined by attaching metal electrodes to both sides of a specimen and applying a voltage between the electrodes until the specimen reaches electrical breakdown. Because breakdown characteristics of the specimen itself should be determined, it is necessary to induce electrical breakdown so that a current passes through the specimen between the electrodes with the shortest distance. For example, when a solid plate specimen is placed between spherical and plate electrodes as shown in Fig. 5.33 and a voltage is applied, a discharge may occur from the spherical electrode. Moreover, a discharge may locally occur in a region with a high electric field intensity because the dielectric constant of the plate specimen is different from that of the medium surrounding the plate specimen.

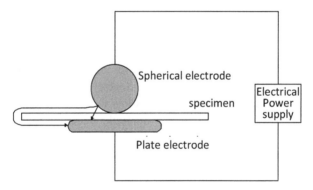

Figure 5.33 Breakdown pathway in sheet specimen.

This discharge may induce electrical breakdown slightly away from the position immediately below the electrode. To prevent such a discharge, the electrodes are embedded in another solid material when the plate specimen to be tested is fabricated. During the test, the plate specimen and electrodes are fully immersed in insulation oil. For film specimens, the section where electrical breakdown is to be induced is thinned and the electrodes are vapor-deposited to cover the thinned section, as shown in Fig. 5.34a. In another method, as shown in Fig. 5.34b, a mask is placed away from a film specimen during the vapor deposition of the electrodes to blur the end of the electrodes. This reduces the intensity of the electric field at the end of the electrodes and prevents the generation of surface discharges [1].

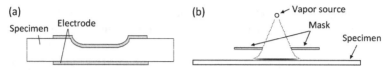

Figure 5.34 Electrode system for film specimen (a) Electrode system for concave specimen (b) Diffused end type electrode.

The withstand voltage depends on the thickness and temperature of the material as well as the waveform, duration, and polarity of the applied voltage. Therefore, these conditions are very important when evaluating the breakdown characteristics of a material and comparing the results among different materials. Testing standards specify the shape of the electrodes, the waveform of the applied voltage, the method of increasing the applied voltage, and other factors.

The mechanism underlying electrical breakdown can be interpreted as follows: (i) the electronic breakdown process that is governed by electrons in a solid, for example, by the loss of energy equilibrium of conduction electrons and the tunneling effect, (ii) the purely thermal breakdown process caused by the change in the balance between the Joule heat and the radiated heat upon the application of a voltage, and (iii) the mechanical breakdown process in which the balance between the stress across the electrodes induced by the application of a voltage and the external force is varied to induce electrical breakdown. These processes are closely related to the electrical breakdown strength, temperature, and thickness of the materials.

5.4.2 How the Short-Term Breakdown Characteristics of Nanocomposite Insulators Change

Let us examine how the short-term breakdown characteristics of nanocomposite insulators change under different conditions. A number of previously reported data on such changes are summarized in Fig. 5.35. As shown in the figure, the breakdown voltage depends on the amount of filler and the waveform of the applied voltage. The ordinate indicates the value obtained by dividing the breakdown voltage of the nanocomposite insulator by that of the base material. The greater this value, the greater the effect of using the insulator in the form of a nanocomposite. Figure 5.35 shows that the breakdown voltage increases when a filler is added to the material but decreases when an excessive amount of filler is added. As shown in Fig. 5.36, the breakdown voltage of the nanocomposite insulators generally increases from that of the base material when a DC voltage is applied; however, it remains unchanged when an impulse voltage is applied and decreases when an AC voltage is applied.

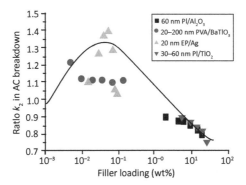

Figure 5.35 Relation between ratio k_2 in AC breakdown voltage and filler loading [2]. The ratio k_2 is defined as breakdown voltage of nanocomposite divided the breakdown voltage of polymer matrix. (▲Breakdown measurements were performed at room temperature (RT) in transformer oil according to the IEC standard. Thickness of samples is about 50 μm. ■ BD measurements were performed at RT in silicon oil. Thickness of samples is 30 +/− 2 μm. The voltage was increased by 1 kV/s. ● Breakdown measurements were performed at 77 K in an open liquid nitrogen bath. Thickness of samples is about 50 μm. The voltage was increased by 500 V/s. ▼ Breakdown measurements were performed at RT in transformer oil according to the IEC standard. Thickness of samples is 40–49 μm. The voltage was increased by 2.5 kV/s). Nanofillers in these papers were all untreated.

The breakdown voltage of nanocomposite insulators also depends on the shape of the electrodes. The effect of using nanocomposites tends to be high for needle–plate electrode systems but tends to be low for parallel-plate electrode systems. Thus, the electrical breakdown characteristics of insulators are not always improved by using them in the form of a nanocomposite. This is because the short-term breakdown characteristics of nanocomposite insulators greatly depend on their fabrication conditions, including the fabrication method, the grain diameter of the filler, the state of aggregation and dispersion, the surface treatment method and procedures, and the water absorption state.

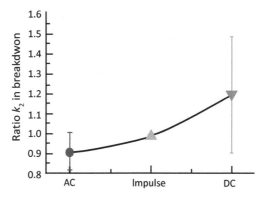

Figure 5.36 Ratio k_2 of nanocomposites in breakdown depend on the applied voltage type [2]. ● AC BD tests were measured with a ramp rate of 500 V/s. The thickness of specimens was ranging from 50 to 500 μm; ▲ Impulse electric strength was measured using a standard impulse of 1/50 μs. The thickness of specimens was ranging from 50 to 500 μm.

5.4.3 Filler State Is Important in Order to Improve Short-Term Breakdown Characteristics

The short-term breakdown characteristics of nanocomposite insulators are improved or deteriorated by various factors, as mentioned above. In the early days of nanocomposite insulators, no established techniques for fabricating nanocomposite materials were available. As a result of the recent improved techniques for fabricating nanocomposites, certain tendencies have been found in the factors affecting the characteristics. Here, we will introduce the electrical breakdown characteristics of nanocomposite insulators by focusing on the state of the nanofiller.

Figure 5.37 shows the dependence of the electrical breakdown strength on the filler size for a nanocomposite fabricated by filling epoxy resin (ER) with boron nitride (BN). The breakdown voltage was measured by applying a DC voltage to a parallel-plate electrode system. The breakdown voltage of the base material is −163 kV/mm, whereas that of the nanocomposite tends to increase with decreasing filler size. A reasonable correlation is observed in the single logarithmic plot, meaning that the breakdown voltage is related to the surface area of the filler.

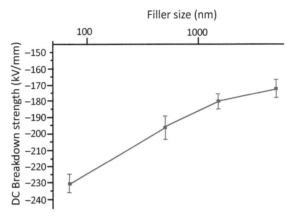

Figure 5.37 DC breakdown strength for BN-ER as function of filler size on a logarithmic scale [3].

Figure 5.38 shows how the electrical breakdown strength of a nanocomposite fabricated by filling polyethylene with a montmorillonite clay nanofiller depends on the dispersion state of the nanofiller. Specimen system A, B, and C are polyethylene, poorly dispersed nanocomposite and good dispersed nanocomposite, respectively. When the nanofiller sufficiently disperses, the breakdown voltage of the nanocomposite is similar to that of the base material. When the nanofiller aggregates, however, the breakdown voltage of the nanocomposite greatly decreases.

Figure 5.39 shows the effect of the surface state of the filler on the electrical breakdown strength. Composites were fabricated by filling epoxy resin with a nano- or microfiller of different surface states (with or without coupling treatment). When a filler without coupling treatment is used, the electrical breakdown strength, particularly that of the nanocomposite, greatly decreases. In the case of using a needle electrode, the growth of an electrical tree affects the breakdown voltage, and therefore the surface treatment of fillers is essential for nanocomposites with a high frequency of collision with the filler.

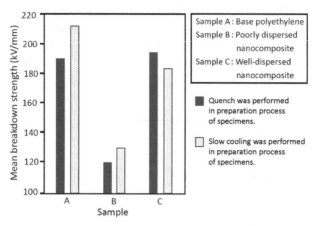

Figure 5.38 Relation between filler dispersion state and breakdown strength [4]. Ramp breakdown data from Systems A, B and C. Data from quenched samples are shown in red, data from speciments crystallized in isothermally at 117°C blue.

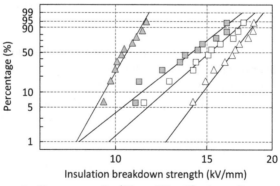

—△— : Nano-composite (12 nm SiO_2 without coupling agent)
—■— : Micro-composite (1.6 μm SiO_2 without coupling agent)
—△— : Nano-composite (12 nm SiO_2 with coupling agent)
—□— : Micro-composite (1.6 μm SiO_2 with coupling agent)

Figure 5.39 Weibull plots of insulation breakdown strength for the nano- and microcomposites with or without a coupling agent [5].

Figure 5.40 shows the effect of water absorption by nanocomposites on the electrical breakdown strength. A nanocomposite was fabricated by filling cross-linked polyethylene with nanosize SiO$_2$. The electrical breakdown strength of the nanocomposite under three conditions and that of the base material were compared to evaluate the effect of water absorption. The electrical breakdown strength of the nanocomposite is high in the dried state; however, it decreases in the wet state to below that of the base material and does not return to the original value even when the nanocomposite is dried again. Therefore, attention should be paid to the water absorption state of nanocomposites.

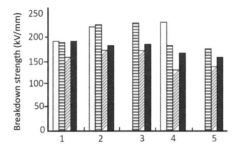

Figure 5.40 The effect of water absorption on breakdown strength (VS: vinyl silane) [6]. *X*-axis: (1) XLPE, (2) XLPE + 5 wt% vs. treated SiO$_2$, (3) XLPE + 5 wt% non-treated SiO$_2$, (4) XLPE + 12.5 wt% vs. treated SiO$_2$, and (5) XLPE +12.5 wt% non-treated SiO$_2$. *Y*-axis: is □, dry; ▤, 75% RH aged; ◇, 100% RH aged; ■ 100% RH aged–vacuum dried.

As above, factors resulting in changes in the short-term breakdown characteristics caused by using insulators in the form of nanocomposites have been examined. The above findings may be used to improve short-term breakdown characteristics.

References

1. Inuishi, Y., Nakajima, T., Kawabe, K., and Ieda, M. (2006). *Dielectric Phenomenological,* 29th ed. (The Institute of Electrical Engineers of Japan, Japan) (in Japanese).
2. Li, S., Yin, G., Chen, G., Li, J., Bai, S., Zhong, L., Zhang, Y., and Lei, Q. (2010). Short-term Breakdown and Long-Term Failure in Nanodielectrcs: A Review, *IEEE Trans. Dielectr. Electr. Insul.,* **5**, pp. 1523–1535.

3. Andritsch, T., Kochetov, R., Gebrekiros, Y. T., Morshuis, P. H. F., and Smit, J. J. (2010). Short Term DC Breakdown Strength in Epoxy Based BN Nano- and Microcomposites, *Proc. IEEE International Conf. Solid Dielectrics (ICSD)*, 1, pp. 179–182.

4. Vaughan, A. S., Green, C. D., Zhang, Y., and Chen, G. (2005). Nanocomposites for High Voltage Applications: Effect of Sample Preparation on AC Breakdown Statistics, *Annual Rept. IEEE Conf. Electrical Insulation and Dielectric Phenomena (CEIDP)*, pp. 732–735.

5. Imai, T., Sawa, F., Ozaki, T., Shimizu, T., Kido, R., Kozako, M., and Tanaka, T. (2005). Evaluation of Insulation Properties of Epoxy Resin with Nano-scale Silica Particles, *Proc. IEEJ International Symposium on Electrical Insulating Materials (ISEIM)*, pp. 239–242.

6. Hui, L., Nelson, J. K., and Schadler, L. S. (2014). Hydrothermal Aging of XLPE/Silica Nanocomposites, *Annual Rept. IEEE Conf. Electrical Insulation and Dielectric Phenomena (CEIDP)*, pp. 30–33.

5.5 Long-Term Dielectric Breakdown (Treeing Breakdown)

Toshikatsu Tanaka

IPS Research Center, Waseda University,
Kitakyushu-shi, Fukuoka 808-0135, Japan

t-tanaka@waseda.jp

A small amount of nanofillers acts predominantly to suppress not only tree initiation but also tree propagation in thick insulation. This effect appears remarkably as long-term treeing breakdown characteristics. This suppressing characteristic can be utilized in practical application. On the other hand, an inverse phenomenon is observed, where short-time tree propagation is accelerated in nanocomposites under a high electric field. This brings about a new insight on how nanofillers interact with transporting high-energy electrons and propagating tree channels.

5.5.1 Treeing Breakdown of Polymers Is Evaluated in Terms of Tree Shapes and *V–t* Characteristics

Treeing is a phenomenon in which when a polymer insulating substance is locally subjected to high electric stress, it suffers from the formation of tree-like dendritic paths and finally from all-path breakdown. This phenomenon is called treeing breakdown. Such dendritic paths are called electrical trees or simply trees. Although treeing breakdown paths cannot necessarily be detected in power apparatuses and cables in service, it is well recognized that this type of dielectric breakdown takes place in thick insulation. For experimental purposes, several types of electrode systems are utilized as shown in Fig. 5.41. In general, a needle electrode is inserted into insulation or molded inside insulation. A counter flat or round electrode is located more than 1 mm away. An electrode system of a metal needle imbedded in a polymer might include a void between them due to the difference in thermal expansion coefficient. A semiconducting polymer (polymer infilled with carbon) electrode is sometimes preferred.

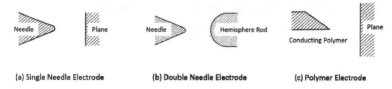

Figure 5.41 Some electrode systems for treeing experiments.

Trees form various shapes that are roughly classified into two types:. tree-like trees and bush-like trees, as shown in Fig. 5.42. In many cases, a tree is initially tree-like, and becomes bush-like as it grows with time. Tree-like trees may be formed over the whole distance (thickness) of a specimen when the electric field at the tip of the needle electrode is high enough to cause dielectric breakdown in a short time. Long-time treeing breakdown includes an incubation period during which nothing appears to happen. A relation between applied voltage and breakdown time is called V–t characteristics. Treeing breakdown is generally evaluated by this relationship. A long-term treeing breakdown under AC voltage includes two time intervals as follows:

(i) Incubation period until tree initiation

(ii) Growth period until breakdown

What takes place during the incubation period peculiar to treeing phenomena? If a specimen has voids around a needle electrode tip, it may be exposed to gaseous partial discharges (PDs) in the voids. The PDs may initiate trees from the region subjected to PDs. In the case of hard materials like epoxy resins, voids may be formed around the needle tip due to repeated Maxwell stress. In that case, trees are formed due to PDs that take place in the voids.

Materials degrade during the incubation period. This concept is widely accepted. In that case, a process of electron injection from the needle electrode is considered to play a predominant role. In the case of polyethylene, as electroluminescence can be observed in the UV (ultraviolet) light region, both electron injection and recombination processes are considered to take place at the same time. UV light is observable during the recombination process. This can cause scission of polymer chains to form radicals, which are easily oxidized when oxygen is present. This is a degradation process by UV light emitted by recombination of injected electrons. Injected electrons may have high enough energy to directly initiate the scission of main polymer chains to cause material degradation. In any case, material degradation brings about reduction in withstanding electric stress, and therefore a tree is generated in a certain period after the application of voltage. This process is illustrated in Fig. 5.43.

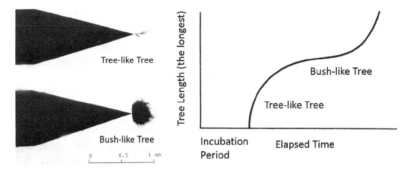

Figure 5.42 Tree-like tree and bush-like tree and their time dependence.

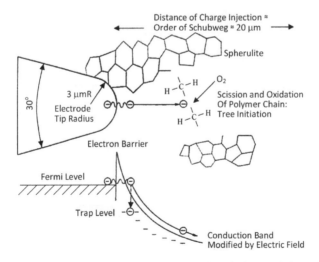

Figure 5.43 Schematic explanation of physical and chemical degradation processes for tree initiation (in the case of polyethylene).

5.5.2 Treeing Breakdown Lifetime Is Enormously Prolonged by Nanofiller Addition

Substantial improvement of treeing breakdown V–t characteristics is expected when polymers are filled with nanofillers to convert them to polymer nanocomposites. Addition of nanofillers into base polymers modifies the tree shapes and retards tree generation and propagation, prolonging lifetime and enhancing dielectric breakdown voltage. Figure 5.44 shows one of such typical V–t characteristics published in the early times of nanocomposite development history [1]. Comparison is made for two kinds of epoxy composites: with titania microfillers and with nanofillers (10 wt% each). This figure demonstrates as a nanofiller effect in which the index n-value (slope) for V–t characteristics is increased, and lifetime is greatly extended. The difference in lifetime becomes enormously larger as applied voltage is lowered. Even 100 times lifetime extension is obtained under an electric field as low as 100 kV/mm. This performance is due more to suppression of tree propagation than to inhibition of tree initiation as described in Section 5.5.3. As practical design stress is generally much lower than 100 kV/mm, this result is considered to be significant, indicating potential practical application of such nanocomposites. Similar characteristics are obtained for epoxy/alumina [2], polyethylene/magnesia [3, 4], and epoxy layered silicate [5, 6] nanocomposites.

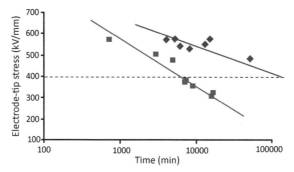

Figure 5.44 Lifetime prolongation by nanofiller addition [1]. ◆, Base resin + 10% nano-TiO$_2$; ■, Base resin + 10% micro-TiO$_2$.

What role would nanofillers play in tree propagation then? They contribute greatly to the suppression of tree propagation.

At least two conceptual mechanisms are proposed. One is that tree channels are forced to form zigzag paths reflected by nanofillers, resulting in their propagation retardation. Another is that tree channels slow down their growth by suppression of their lateral expansion caused by nanofillers. Strong evidence is shown for the latter in Fig. 5.45. SEM photos are taken on inner surfaces of tree channels formed in base epoxy resins and derived nanocomposites. The inner surfaces of tree channels are smooth in the base epoxy resins, while the inner surfaces are rough with some deposits in the nanocomposites. Inner surfaces of tree channels are subjected to internal PDs during propagation of tree channels inside which PDs are taking place. Deposits are considered to be aggregates of nanofillers. They are released on the internal surfaces due to erosion of nanocomposites caused by PD attack. Nanocomposites are eroded by PDs to release organic epoxy as vapor and inorganic nanofillers as deposits. That is to say, the inner surfaces of the base epoxy resins are eroded smoothly by PDs, but those of nanocomposites exhibit strong resistance to PDs due to nanofillers included inside. For this reason, treeing lifetime is extended in nanocomposites. Nanocomposites have naturally strong PD resistance, and this performance appears in treeing phenomena to extend their lifetime.

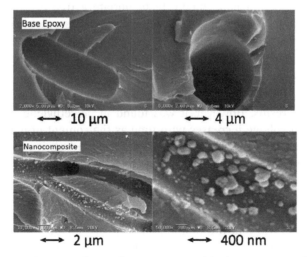

Figure 5.45 Inner surfaces of trees generated in base epoxy and epoxy nanocomposite [2].

5.5.3 What Role Do Nanofillers Play in Tree Initiation?

It has been clarified that treeing breakdown lifetime is substantially prolonged by the mechanism in which nanofillers suppress erosion inside tree channels caused by PDs. Furthermore, nanofillers exert significant influence on behaviors of electrons associated with tree initiation. That is to say, nanofillers will trap and scatter electrons transported inside nanocomposites and force propagating trees to zigzag and branch.

Tree generated in the early stage is called an initial tree. Nanofillers retard the generation of initial trees with 10 to 100 μm in length [7, 8]. Figure 5.46 demonstrates such retarding effects on tree initiation. From this figure, the following conclusions can be drawn:

(i) Comparison of tree initiation time between base epoxy resins and nanocomposites to initiate a tree of 10 μm in length: Initiation time is increased by the addition of nanofillers, and the index n in V–t characteristics is increased from 9.5 to 14. This increase in the index n means that tree initiation time is remarkably extended under low or moderate electric stress. It can be said that nanofillers bring about a lifetime shift effect and an index n increase effect.

(ii) Comparison of tree initiation time between base epoxy resins and nanocomposite to initiate a tree of 100 μm in length: Qualitatively the same two effects as stated above are effective. The index n increases from 10 to 16 by the index n increase effect in this case.

No clear evidence has yet been identified on whether or not tree initiation time is shortened by the incorporation of nanofillers in epoxy resins. However, it was found in polyethylene that tree initiation voltage increases with time as the content of nanofillers (MgO in this case) increases [4]. From this finding, it can be said that nanofillers play some role on tree initiation. It is proposed as a mechanism for this behavior that electrons injected from a needle electrode tip might be trapped on nanofillers to make it difficult for electron avalanche to happen. For that reason, nanofillers will retard tree initiation time and enhance breakdown voltage. Another mechanism is that electrons are scattered and decelerated by the electric field created by nanofillers, making it difficult for electron avalanche to take place. Figure 5.47 shows an interesting SEM photo in which a minute tree is generated from a bulky tree channel in epoxy nanocomposites. This gives some insight into the inner structures of tree channels in nanocomposites. This minute tree is approximately 1 µm in length extending from a big tree channel, and is considered to be similar to the initial tree generated from a needle electrode tip. The minute tree is hindered even in the region shorter than 1 µm and is forced to form zigzag paths and branches.

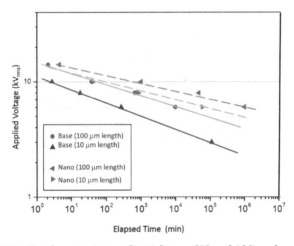

Figure 5.46 V–t characteristics of initial trees (10 and 100 µm length).

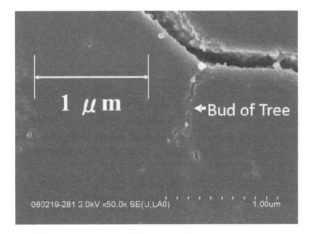

Figure 5.47 Tiny tree generated from a bulky tree trunk.

5.5.4 A Crossover Phenomenon Appears: Tree Growth vs. Voltage

Nanofillers act to suppress or decelerate tree propagation under the low electric field condition. However, an apparently strange phenomenon is observed: Nanofillers act inversely to accelerate tree propagation under a high electric stress. This strange inversed phenomenon is called "the crossover phenomenon in tree propagation," as shown in Fig. 5.48 [2]. This demonstrates the tree length vs. applied voltage characteristics obtained using a needle-to-plane electrode system (spaced 3 mm between a metal needle of 1 mm diameter and 5 µm tip diameter). It is clear that tree length is shorter in nanocomposites at voltages below 17 kV$_{rms}$ than in their base epoxy resins, while it is longer above this voltage in nanocomposites than in base resins. This finding clearly shows that the speed of electron transit or tree channel tip propagation is increased by the incorporation of nanofillers under the high electric stress condition.

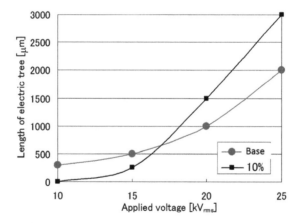

Figure 5.48 Cross-over phenomenon: Relation between tree length 4 min after voltage application and applied voltage in base epoxy and 10 wt% alumina nanofiller added nanocomposite [2].

It should be noted that the reach of minute tree channels to the counter plane electrode does not necessarily mean dielectric breakdown of a specimen. High conducting paths should be formed inside laterally expanded tree channels for total breakdown, like return strokes in long-distance gaseous discharges. Erosion of inner surfaces of tree channels takes place due to PDs. In the case of nanocomposites, nanofillers act to retard erosion and thereby extend dielectric breakdown time.

5.5.5 How Do Nanofillers Act for Tree Growing Processes?

Treeing is governed by two processes: initiation and propagation. Nanofillers bring about enhancement of tree initiation voltage and prolongation of tree initiation time. These are the fundamental phenomena that will give much insight to how nanofillers are related to tree initiation. A tree generated from a metal needle tip and another tree branched from the side of a bigger tree are all considered to have the same mechanisms. This can be understood from its inherent fractal nature. Remaining problems are to clarify how transiting electrons and propagating minute trees electrically and physically interact with nanofillers.

Figure 5.49 illustrates an image constructed based on the knowledge obtained to date on how tree propagation is influenced by nanofillers [7]. Tree propagation is hindered by nanofillers under the low electric field condition. That is, under the low electric field condition, a tree collides with nanofillers and grows a in zigzag pattern when it is small, and it is resisted strongly by nanofillers in a process in which it suffers from PD erosion and its inner channel diameter expands. Under the high electric stress condition, propagation of electrons and tree channels are accelerated. These two processes together explains why the "crossover phenomenon" takes place. How they are accelerated remains yet unsolved. However, it might be taken into consideration that nanofillers induce negative charge up to the Debye shielding length, which naturally builds up the electric field to accelerate electrons.

(i) Figure 5.49 (a) Low Voltage Applied: Two trees are illustrated as the same length for (a-1) and (a-2), when time is shorter in (a-1) than in (a-2). A tree initiation is actually suppressed in the case of (a-2).

(ii) Figure 5.49 (b) High Voltage Applied: Two trees are illustrated in the same time scheme. Channels are accelerated in the case of (b-2). Dielectric breakdown takes place immediately in the case of (b-1) when tree channels reach the counter electrode. Breakdown is delayed with the time required to enlarge the diameter of channels by the PD erosion process.

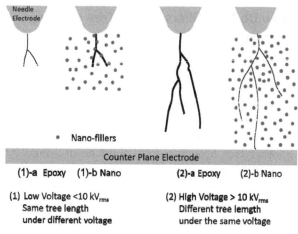

Figure 5.49 Effects of nanofillers on tree propagation.

References

1. Nelson, J. K., and Hu, H. (2004). The Impact of Nanocomposite Formulations on Electrical Voltage Endurance, *Proc. IEEE ICSD*, No. 7P-10, pp. 832–835.
2. Tanaka, T., Matsunawa, A., Ohki, Y., Kozako, M., Kohtoh, M., and Okabe, S. (2006). Treeing Phenomena in Epoxy/Alumina Nanocomposite and Interpretation by a Multi-Core Model, *IEEJ Trans. Fundamentals Mater.*, **126**(11), pp. 1128–1135.
3. Tanaka, T., Nose, A., Ohki, Y., and Murata, M. (2006). PD Resistance valuation of LDPE/MgO Nanocomposite by a Rod-to-Plane Electrode System, *Proc. IEEE ICPADM*, **1**(L–1), pp. 319–322.
4. Kurnianto, R., Murakami, Y., Nagao, M., Hozumi, N., and Murata, Y. (2007). Treeing Breakdown in Inorganic-Filler/LDPE Nano-Composite Material, *IEEJ Trans. Fundamentals Mater.*, **127**(1), pp. 29–34.
5. Tanaka, T., Yazawa, T., Ohki, Y., Ochi, M., Harada, H., and Imai, T. (2007). Frequency Accelerated Partial Discharge Resistance of Epoxy/Clay Nanocomposite Prepared by Newly Developed Organic Modification and Solubilization Methods, *Proc. IEEE ICSD*, No. D1-8, pp. 407–410.

6. Raetzke, S., Ohki, Y., Imai, T., Tanaka, T., and Kindersberger, J. (2009). Tree Initiation Characteristics of Epoxy Resin and Epoxy/ Clay Nanocomoposite, *IEEE Trans. Dielectr. Electr. Insul.*, **16**(5), pp. 1473–1480.

7. Tanaka, T. (2011). Comprehensive Understanding of Treeing *V–t* Characteristics of Epoxy Nanocomposites, *Proc. ISH*, No. E-008, pp. 1–6.

8. Tanaka, T., Iizuka, T., and Wu, J. (2011). Generation Time and Morphology of Infancy Trees in Epoxy/Silica Nanocomposite, *Proc. IEEJ ISEIM*, No. A-2, pp. 5–8.

5.6 Insulation Degradation (Material Degradation Due to Partial Discharges)

Toshikatsu Tanaka

IPS Research Center, Waseda University,
Kitakyushu-shi, Fukuoka 808-0135, Japan

t-tanaka@waseda.jp

Positive modification of polymer characteristics due to the addition of nanofillers appears markedly in PD (partial discharge) resistance. Interestingly, various kinds of performance of nanofillers are exhibited in their PD resistance processes. Improvement of PD resistance using these characteristics will lead to the broadened use of nanocomposites in electric power apparatuses and electronic devices.

5.6.1 PD Resistance of Polymers Valuated in Terms of Erosion Phenomena

Partial discharges (PDs) are similar to corona discharges caused by local or partial breakdown of atmospheric air that can be seen at night as light emission around overhead power transmission lines. This partial breakdown can be reproduced if high voltage is applied to a sharp metal electrode tip in the air. When a compound consisting of solid and gas insulation is subjected to voltage, it generates gaseous discharges since the air has lower breakdown voltage than the solid. These are called PDs in the field of solid insulation. They are generally classified into two discharge forms: the Townsend discharge type (discharges in a whole space) and the streamer discharge type (tree-like discharges). Such discharges can be generally detected as electric pulses. Since solid insulation is subjected to such discharges for a long time, it suffers from material degradation and finally from ultimate breakdown. This is a process of PD degradation.

PDs also take place in liquids to induce evaporation and produce gaseous discharges, which have phenomenological resemblance to solid-gas compound insulation. Insulation degradation due to PDs has long been investigated in various insulations in the power sector. Among them are SL (screened lead sheathed) cables based on oil impregnated insulation and power generator winding insulation based on mica-asphalt compounds and mica-epoxy compounds that are both subjected to PDs in service. No PDs take place normally in service in OF cables (pressurized oil-filled cables), POF cables (pipe-type OF cables), or oil-immersed power transformers. PDs are not likely to occur in XLPE cables (cross-linked polyethylene insulated cables) and molded power transformers. PDs are an important target for research, because organic polymer materials in use are easily degraded when subjected to PDs constantly for the former under normal conditions or occasionally for the latter under abnormal conditions.

Recently a new PD problem has arisen where inverter drive circuits are utilized to drive motors. Fast rise time electric pulses called inverter surges are generated from PWM-inverter-fed circuits to bring about PD pulses which degrade motor winding insulation. Therefore, even in the microelectronics sector, PDs now attract much attention. PDs take place for example in (i) devices used for back lighting of liquid crystal TV sets such as inverter transformers, high voltage printed wiring boards (PWBs), connectors, connecting cables, (ii) high voltage switching power sources for projector light sources, strobe circuits of digital cameras and negative ion generators, and (iii) inverter-fed motors used for air conditioners, washing machines and refrigerators.

Paschen's law ($V = f(pd)$) determines the dielectric breakdown strength (gaseous discharge inception voltage) of the air with initial electrons in electrical insulation systems. This is represented by a function of pd, where d and p are the air gap and the air pressure, respectively.

Figure 5.50 shows several electrode systems devised to evaluate PD resistance. PDs include (i) surface discharges and (ii) internal discharge. Figures 5.50a–d belong to the former, while Figs. 5.50e–g belong to the latter. Surface discharges take place in the open air, and therefore the kind of gases used can be kept for a whole pre-determined experimental time span. Since PD resistance experiments are generally carried out under atmospheric conditions, oxidation reaction processes are involved in the surface discharges. Since internal discharges take place in an enclosed space, oxygen in the air is consumed by oxidation of the insulation surface. Constituent gases chemically and physically change with time from oxygen-rich to nitrogen-rich gases including vaporized solid insulation. The inside pressure also changes. Therefore, characteristics of gaseous discharges are also modified depending on the respective ratio of the constituent gases. This modification further influences the process of material degradation.

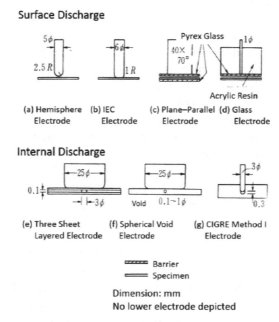

Figure 5.50 Various types of electrode systems for PD degradation tests.

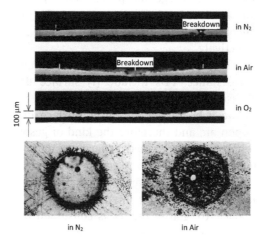

Figure 5.51 Uniform erosion by void PDs and pit formation grown PDs.

When insulating materials are subjected to PDs in the air, they are eroded directly by PDs and are also physically and chemically degraded indirectly by discharge-originated products such as activated oxygen, ozone, oxidized nitrogen, and acids. That is, (i) electrons, positive and negative ions, and excitons, (ii) heat and ultraviolet light, and (iii) high energy ions are produced in the discharge space. Materials are PD-degraded by one or multi-factors including oxidation reactions, ion bombardment and thermal decomposition. Synergy effects may also work. Oxygen might be consumed into polymers to form carbonyl groups. In the case of polyethylene, PDs induce oxalic acid if no further air is supplied, and nitric acid if water exists. Chemical reactions occur in oxygen discharges to cause oxidation degradation, while physical attacks take place in nitrogen discharges to cause ion bombardment degradation. Oxalic acid and nitric acid are also detected in epoxy resins. The oxalic acid must be a thermal degradation product of the phthalic anhydride contained in the hardener. Figure 5.51 shows some profiles of degradation of polyethylene clarified by the PD degradation experiments using specimens consisting of three-layer sheets. Oxygen discharges induce homogeneous erosion, nitrogen discharges create pits, and aerial discharges produce something like middle patterns.

5.6.2 PD Resistance of Polymers Is Drastically Increased Due to Nanofiller Inclusion

It was experimentally found first that nanocomposites are highly resistant to PDs. When conventional resins such as epoxy and polyethylene include nanofillers of several wt% as additives, they are enormously improved in PD erosion resistance, as shown in Fig. 5.52. PD resistance is increased up to 10 times, i.e. the inverse erosion index (200 μm/20 μm) in this case. This value is obtained using an electrode system as shown in Fig. 5.53. This is a modified electrode system out of Fig. 5.50a with the difference being that it has a 0.2 mm air gap. Experiments are also carried out using an IEC (b) electrode as shown in Fig. 5.50b. The erosion depth depends on the distance from the electrode edge in this case, and some other parameter like the surface roughness is an index of PD resistance.

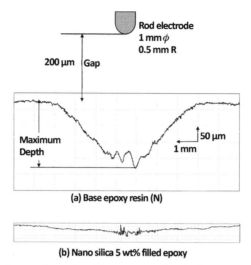

Figure 5.52 Comparison in PD erosion depth between base epoxy and epoxy/silica nanocomposite: 60 Hz equivalent time 1440 h at 4 kV$_{rms}$.

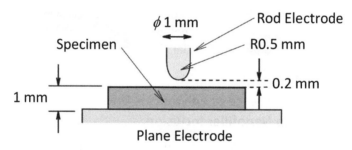

Figure 5.53 Rod-plane electrode system used for PD resistance evaluation tests.

PD resistance varies from filler to filler as shown in Fig. 5.54. This result is beneficial for filler selection. But it should be noted that the result is influenced not only by the kind of fillers but also by nanocomposite preparation procedures like the quality of interfaces formed between fillers and polymers and the difference in curing conditions. In addition to epoxy, similar performance is ascertained with other materials like polyethylene and polypropylene. Therefore, it is a general characteristic that nanocomposites are more PD-resistant than their base polymers.

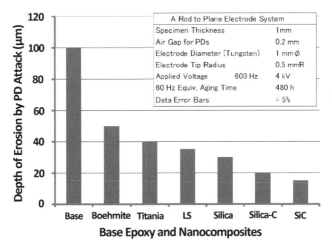

Figure 5.54 Comparison of PD erosion depth among various kinds of nanocomposites.

Knowledge accumulated on PD resistance of nanocomposites is as follows [1–3]:

(i) Nanofiller addition is effective for conventional polymers for electrical insulation purposes.
(ii) Filler diameter for effectiveness should be as small as possible in the range of 10 nm to 1000 nm.
(iii) Filler surface modification using methods such as a silane coupling is recommended.
(iv) Hydrophilic fillers are more appropriate than hydrophobic fillers.

Similar results are obtained with epoxy/alumina [4], epoxy/alumina, titania [5, 6], epoxy/silica, alumina [7, 8], epoxy/silicon carbide (SiC) [9], polyethylene/magnesia [10], polypropylene/polyaniline/acrylic acid [11].

5.6.3 What Are Mechanisms for PD Erosion in Nanocomposites?

When nanofillers are added to conventional polymer insulating materials, they are converted to nanocomposites with stronger PD resistance. It is of interest to know what role nanofillers play in mechanisms associated with strong resistance of nanocomposites.

PD degradation processes are described for single materials in Section 5.6.1. Mechanistic analysis is necessary for multi-component materials in the case of nanocomposites. Mechanisms clarified so far are as follows:

(i) Polymers are divided three-dimensionally by nanofillers with strong PD resistance.
(ii) Polymers are bonded tightly with nanofillers.
(iii) Polymers are covered with nanofiller deposited layers after PD attack.

The above three phenomena are represented by (1) nanosegmentation, (2) interfacial reinforcement, and (3) nanofiller accumulation. Such mechanisms as nanofiller addition effects are illustrated in Fig. 5.55.

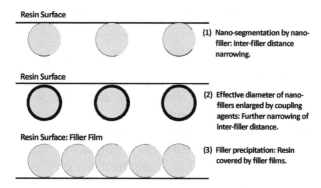

Figure 5.55 Schematic explanation of roles of nanofillers to improve PD resistance.

First, Fig. 5.55a shows that the resin of interest is three-dimensionally divided in the nanometer scale by nanofillers. This structure is considered to play a big role in determining PD resistance. For instance, inter-distances between two neighboring nanofillers are in the range of 40 nm if nanofillers of 20 nm diameter are added by the small amount of 5 wt%. The resin is as a result meshed three-dimensionally into 40 nm. Since nanofillers as inorganic materials are almost immune to PDs in general, only parts of the resin of several tens of nanometers are exposed to PDs to be degraded. As a result, nanocomposites as a total composite material become PD resistant. In addition, electric stress is lessened near the parts of the resin, since electric field is intensified on the nanofillers because the nanofillers have larger permittivity than the resin. This will also assist nanocomposites so that they might become stronger in PD resistance. Second, nanofiller diameters are apparently increased if interfaces with thickness are formed by coupling agents, as shown in Fig. 5.55b. Parts of the resin with lower PD resistance become narrow, resulting in an increase in PD resistance of all the composites. Third, nanofillers are deposited on the surface of the nanocomposite of interest to form PD-resistant nanofiller layers, since the resin-like epoxy is vaporized, as shown in Fig. 5.55c. Under this condition, PDs trend to diminish. In an extreme case, inorganic layers are formed via a sintering process caused by gaseous discharges.

Furthermore, it should be noted that the excellent performance of nanocomposites might be worsened if sufficient care is not taken in fabrication processes. PD erosion proceeds easily if interfaces are not good enough to prevent voids in the nanocomposites. If the nanocomposites include voids, they are subject to water permeation and degrade easily.

Reference

1. Kozako, M., Kuge, S., Imai, T., Ozaki, T., Shimizu, T., and Tanaka, T. (2005). Surface Erosion Due to Partial Discharges on Several Kinds of Epoxy Nanocomposites, *Annual Rept. IEEE-CEIDP*, No. 2C-5, pp. 162–165.

2. Tanaka, T., Ohki, Y., Shimizu, T., and Okabe, S. (2006). Superiority in Partial Discharge Resistance of Several Polymer Nanocomposites, *CIGRE Paper*, D1-303, p. 8.

3. Tanaka, T., Kuge, S., Kozako, M., Imai, T., Ozaki, T., and Shimizu, T. (2006), Nano Effects on PD Endurance of Epoxy Nanocomposites, *Proc. ICEE*, p. 4.

4. Kozako, M., Ohki, Y., Kohtoh, M., Okabe, S., and Tanaka, T. (2006). Preparation and Various Characteristics of Epoxy/Alumina Nanocomposites, *IEEJ Trans. Fundamentals Mater.*, **126**(11), pp. 1121–1127.

5. Tanaka, T., and Iizuka, T. (2010). Generic PD Resistance Characteristics of Polymer (Epoxy) Nanocomposite, *Annual Rept. IEEE-Conf. Electr. Insul. Dielectr. Phenomena*, No. 7A-1, pp. 518–521.

6. Maity, P., Basu, S., Parameswaran, V., and Gupta, N. (2008). Degradation of Polymer Dielectrics with Nanometric Metal-Oxide Fillers due to Surface Discharges, *IEEE Trans. Dielectr. Electr. Insul.*, **17**(1), pp. 52–62.

7. Maity, P., Kasisomayajula, S. V., Parameswaran, V., Basu, S., and Gupta, N. (2008). Improvement in Surface Degradation Properties of Polymer Composites due to Pre-Processed Nanometric Alumina Fillers, *IEEE Trans. Dielectr. Electr. Insul.*, **17**(1), pp. 63–72.

8. Preetha, P., Alapati, S., Singha, S., Venkatesulu, B., and Thomas, M. J. (2008). Electrical Discharge Resistant Characteristics of Epoxy Nanocomposites, *IEEE Annual Rept. CEIDP*, No. 8-4, pp. 718–721.

9. Tanaka, T., Matsuo, Y., and Uchida, K. (2008). Partial Discharge Endurance of Epoxy/SiC Nanocomposite, *Annual Rept. IEEE-CEIDP*, No. 1-1, pp. 13–16.

10. Tanaka, T., Nose, A., Ohki, Y., and Murata, Y. (2006). PD Resistance valuation of LDPE/MgO Nanocomposite by a Rod-to-Plane Electrode System, *Proc. IEEE ICPADM*, **1**(L-1), pp. 319–322.

11. Takala, M., Sallinen, T., Nevalainen, P., Pelto, J., and Kannus, K. (2009). Surface Degradation of Nanostructured Polypropylene Compounds Caused by Partial Discharges, *Proc. IEEE ISEI*, No. S3-2, pp. 205–208.

5.7 Insulation Degradation (Material Degradation Due to Water Trees)

Masayuki Nagao[a], Naoki Hayakawa[b] and Muneaki Kurimoto[b]

[a]Department of Electrical and Electronic Information Engineering,
Toyohashi University of Technology,
Toyohashi-shi, Aichi 441-8580, Japan
[b]Department of Electrical Engineering and Computer Science,
Nagoya University, Nagoya-shi, Aichi 464-8603, Japan

nagao@tut.jp, nhayakawa@nuee.nagoya-u.ac.jp, kurimoto@nuee.nagoya-u.ac.jp

It is known that the addition of a small amount of nanofiller to a polymer can suppress the development of water trees. In particular, polymer lifetime can be markedly improved as a result of inhibiting the long-term deterioration by water trees. Interestingly, the interaction between water and nanofillers is related to the suppression effect. In this section, the mechanism behind the suppression of the development of water trees by the addition of nanofillers is explained.

5.7.1 Water Trees Are Generated When Polymers Are Subjected to Both Water and Electric Field

Dendritic deterioration marks are found in polyethylene-insulated wires (used for underwater motors) and cross-linked polyethylene-insulated cables (called XLPE cables or CV cables). They have a shape similar to that of electrical trees and are called water trees. Water trees are generated when the following two conditions are satisfied.

(i) Polymers absorb water.
(ii) A high electric field is applied to the polymers.

Water trees further form continuous channels similarly to electrical trees. Water trees are generated and develop upon a long-term application of even an extremely low voltage. Therefore, from an engineering viewpoint, it is possible to alleviate this problem by preventing the invasion of water into an insulator or by preventing the formation of local regions with a high electric field. In practice, cables with good insulation performance are realized by supplying a water-impervious layer and removing the defects that cause the formation of local regions with a high electric field, such as voids and foreign substances in the insulators and protrusions from the semiconductor layer as much as possible.

The shape of water trees depends on the position of their generation in a cable. Microphotographs of water trees generated in a polymer insulator of an XLPE cable are shown in Fig. 5.56. Figure 5.56a shows a dendritic water tree. Such a water tree is generated from defects such as a conductive protrusion at the interface between the polymer and a semiconductor layer. Figure 5.56b shows a water tree that develops in the shape of a bow tie. This water tree is generated when defects such as voids and foreign substances exist in the polymer. Such a water tree is particularly called a bow-tie tree.

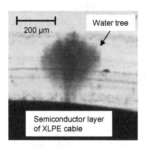
(a) Water tree generated from conductive protrusion

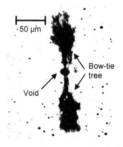
(b) Water tree generated from void exist in the polymer [1]

Figure 5.56 Water trees generated in a polymer insulator of an XLPE cable. Figure (b): © 1990 IEEE. Reprinted, with permission, from Steennis, E. E., and Kreuger, E. H. (1990). Water Treeing in Polyethylene Cables, *IEEE Trans. Dielectr. Electr. Insul.*, **25**(5), pp. 989–1028.

The results of scanning electron microscopy (SEM) reveal that a water tree is a collection of microvoids and paths connecting them. These components are formed as a result of the cleavage of the molecular chains of polymers. The mechanism of water tree development consists of the following two processes, although it has not been fully clarified [2, 3]:

(i) Water accumulation in regions with a high electric field
(ii) The generation and development of a water tree from the above regions

Figure 5.57 shows the conditions under which water trees are generated [4]. Namely, the conditions such as the applied voltage, frequency of voltage, type of material, oxidation resistance, temperature, water quality, amount of water, and mechanical strain are considered to affect the generation of water trees.

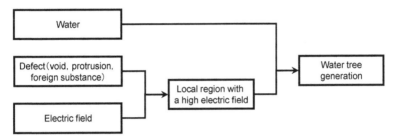

Figure 5.57 Conditions under which water trees are generated.

Water electrodes are frequently used in the research of water trees to directly apply voltage to water, as shown in Fig. 5.58. This is because they can generate water trees in a short time, although the time required to accumulate water for the generation of water trees is not included. A water-soluble ionic substance (e.g., NaCl) is added to control the conductivity of water. A needlelike dent, a void, or a foreign substance is introduced to form a local region with a high electric field as the starting point of a water tree.

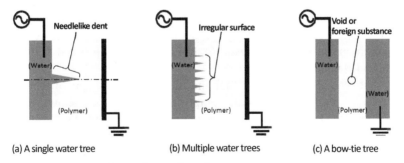

Figure 5.58 Test electrodes used to observe water tree.

Figure 5.58a shows a needlelike dent on the surface of a specimen used to observe a single water tree in detail. The needlelike dent is made, for example, by inserting a needle with a micrometer-order-diameter tip into the specimen surface and then withdrawing it.

Figure 5.58b shows the irregular surface of a specimen used to observe multiple water trees effectively. The irregular surface is made, for example, by rubbing sand paper over the surface of the specimen.

Figure 5.58c shows the specimen with a void or a foreign substance inserted to observe a bow-tie tree. On both sides of the specimen are water electrodes.

5.7.2 Progress of Water Trees Is Suppressed by Nanofiller Addition

The suppression of the development of water trees in nanocomposites has been clarified, although there have been no reports on the suppression of the generation of water trees in nanocomposites, to the best of our knowledge. Figure 5.59 shows the Weibull distribution of the water-tree length in a nanocomposite obtained by adding 2.5 wt% nanosilica to low-density polyethylene (LDPE) and LDPE alone [5]. The electrode shown in Fig. 5.58b was used in the experiment. The length of the water tree formed in the nanocomposite obtained by applying an AC voltage of 5 kV for 45 days is approximately half that of the water tree formed in the LDPE alone. There is no evidence that the generation of water trees is delayed in the nanocomposite. However, the development of water trees is clearly suppressed in the nanocomposite, as explained later. In addition, the development of water trees is confirmed to be markedly suppressed using nanofillers in XLPE/magnesia [6], LDPE/ magnesia [6], and XLPE/ silica [7] nanocomposites.

It was also confirmed that the rate of development of water trees is reduced using nanocomposites. Figure 5.60 shows the water-tree length in the nanocomposite obtained by adding magnesia nanofillers to LDPE [6]. Water trees form upon applying a voltage with a commercial frequency of AC 5 kV using the electrode shown in Fig. 5.58b. The water-tree length decreases with increasing amount of nanofiller (from 0 parts per hundred of resin (phr) to 5 phr). In particular, the gradient of water-tree length becomes gentle with time, indicating that the nanofiller suppresses the development of water trees.

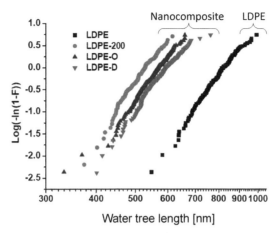

Figure 5.59 Weibull distribution of the water-tree length in nanocomposites obtained by adding 2.5 wt% nanosilica to low-density polyethylene (LDPE) and LDPE. © 2009 IEEE. Reprinted, with permission, from Huang, X., Ma, Z., Jiang, P., Kim, C., Liu, F., Wang, G., and Zhang, J. (2009). Influence of Silica Nanoparticle Surface Treatments on the Water Treeing Characteristics of Low Density Polyethylene, *Proc. IEEE ICPADM*, No. H-7, pp. 757–760.

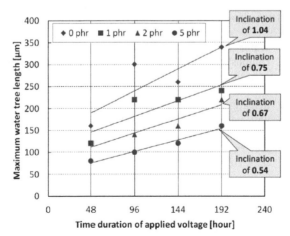

Figure 5.60 Water-tree length in the nanocomposite obtained by adding magnesia nanofillers to LDPE [6].

5.7.3 What Role Do Nanofillers Play to Suppress Water Tree Progress?

The nanofiller interface plays an important role in suppressing the development of water trees. Figure 5.61a shows the water-tree lengths of the nanocomposites obtained by adding hydrophilically treated nanosilica (hydrophilic nanosilica) or hydrophobically treated nanosilica (hydrophobic nanosilica) to linear LLDPE [8]. The water trees begin to develop upon applying an AC voltage of 5 kV for 45 days using the electrode shown in Fig. 5.58b. The water tree in the nanocomposite obtained by adding hydrophilic nanosilica is shorter than that in the nanocomposite obtained by adding hydrophobic nanosilica.

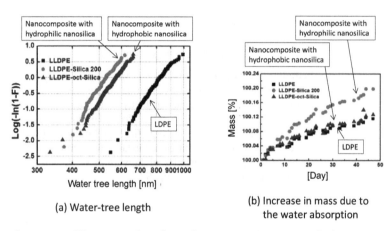

(a) Water-tree length

(b) Increase in mass due to the water absorption

Figure 5.61 Water-tree length and increase in mass of the nanocomposites obtained by adding hydrophilically treated nanosilica (hydrophilic nanosilica) or hydrophobically treated nanosilica (hydrophobic nanosilica) to linear low-density polyethylene (LLDPE). © 2010 IEEE. Reprinted, with permission, from Huang, X., Liu, F., and Jiang, P. (2010). Effect of Nanoparticle Surface Treatment on Morphology, Electrical and Water Treeing Behavior of LLDPE Composites, *IEEE Trans. Dielectr. Electr. Insul.*, **17**(6), pp. 1697–1704.

Figure 5.61b shows the increase in mass due to the water absorption of the above two nanocomposites. Only the mass of the nanocomposite obtained by adding a hydrophilic nanofiller (nanosilica) increases. This phenomenon can be explained by a model in which a water shell is formed when a hydrophilic nanofiller interface absorbs water [9]. From the mechanism based on the water shell effect, the hydrophilic nanofiller interface traps water, prevents the accumulation of water required for the development of water trees, and hence suppresses the development of water trees. The development of water trees is also suppressed by adding a hydrophobic nanofiller, as shown in Fig. 5.61a, although the adsorption of water on this hydrophobic nanofiller is not observed, as shown in Fig. 5.61b. Therefore, it is considered that there are causes other than water absorption behind the suppression of the development of water trees by nanofillers.

The micrographs of the water tree in XLPE alone (a) and the water tree in the nanocomposite obtained by adding 5 wt% silica nanofillers to XLPE (b) are shown in Fig. 5.62 [7]. The water trees begin to develop upon applying an AC voltage of 5 kV for 6 days using the electrode shown in Fig. 5.58a. The water tree in the nanocomposite spreads in the shape of a fan with respect to the direction of the electric field, and extends wider than that in XLPE alone. From this finding, the sterical division of the polymer by the filler is effective, as observed in electrical trees and partial discharge deterioration, leading to the development of zigzag water trees. This phenomenon is mainly observed in the nanocomposite using XLPE as a base polymer.

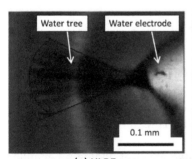
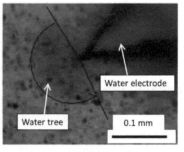

(a) XLPE   (b) Nanocomposite (nanosilica: 5 wt%)

Figure 5.62 Micrographs of the water tree in XLPE alone and the water tree in the nanocomposite obtained by adding 5 wt% silica nanofillers to XLPE. © 2009 IEEE. Reprinted, with permission, from Hui, L., Smith, R., Nelson, J. K., and Schadle, L. S. (2009). Electrochemical Treeing in XLPE/Silica Nanocomposites, *Annual Rept. IEEE CEIDP*, No. 6-2, pp. 111–114.

Figure 5.63 shows the dependence of the maximum water-tree length on temperature in the nanocomposite obtained by adding magnesia nanofillers to LDPE [10]. When the amount of added nanofiller is relatively small (1 or 2 phr), the effect of suppressing the development of water trees that is observed at low temperatures tends to be weak at a high temperature (60°C). However, this effect is maintained at 60°C when the amount of added nanofiller is 5 phr. Researchers found that the volume of spherulite in polyethylene increases and that of the amorphous phase decreases upon the addition of nanofillers [8]. This contributes to the suppression of the development of water trees. It is considered that with increasing amount of added nanofiller, the volume of the amorphous phase with a low mechanical strength decreases in the polyethylene, and the cleavage of the polymer chains, which develops water trees, is suppressed.

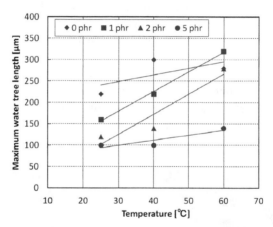

Figure 5.63 Dependence of the maximum water-tree length on temperature in the nanocomposite obtained by adding magnesia nanofillers to LDPE [10].

Although the investigation of the effect of adding a nanofiller on the suppression of the generation of water trees will progress through future studies, the effect on the suppression of the development of water trees was demonstrated as explained above. This effect is considered to be expressed in any combination of polyethylene (LLDPE, LDPE, and XLPE) and nanofiller material. The mechanism of the suppression of the development of water trees by nanofillers largely depends on the type of base polymer.

The suppression mechanism for LDPE is summarized in Fig. 5.64 [10]. The nanofiller interface is considered to trap water as if it were a water shell, during the development of water trees in LDPE. Because of this, the accumulation of water required for the development of water trees is suppressed in the local regions with a high electric field. In addition, the volume of spherulite in LDPE increases with the addition of a nanofiller, leading to a decrease in the volume of the amorphous phase with a low strength. Thus, the cleavage of polymer chains tends not to occur, further suppressing the development of water trees. Note that the development of water trees in the direction of the electric field is suppressed in the nanocomposite containing a base polymer with a high crosslink density, such as XLPE, because the water trees develop in the shape of a fan.

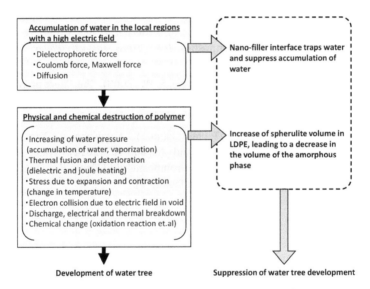

Figure 5.64 Mechanism of the suppression of the development of water trees by nanofillers [10].

References

1. Steennis, E. E., and Kreuger, E. H. (1990). Water Treeing in Polyethylene Cables, *IEEE Trans. Dielectr. Electr. Insul.*, **25**(5), pp. 989–1028.
2. Crine, J. P. (1998). Electrical, Chemical and Mechanical Processes in Water Treeing, *IEEE Trans. Dielectr. Electr. Insul.*, **5**(5), pp. 681–694.

3. Patsch, R. (1992). Electrical and Water Treeing: A Chairman's View, *IEEE Trans. Dielectr. Electr. Insul.*, **27**(3), pp. 532–542.

4. Hayami, T. (1986). *CV Cable*, Ohmsha, pp. 55 (in Japanese).

5. Huang, X., Ma, Z., Jiang, P., Kim, C., Liu, F., Wang, G., and Zhang, J. (2009). Influence of Silica Nanoparticle Surface Treatments on the Water Treeing Characteristics of Low Density Polyethylene, *Proc. IEEE ICPADM*, No. H-7, pp. 757–760.

6. Nagao, M., Watanabe, S., Murakami, Y., Murata, Y., Sekiguchi, Y., and Goshowaki, M. (2008). Water Tree Retardation of MgO/LDPE and MgO/XLPE Nanocomposites, *Proc. IEEJ ISEIM*, No. P2-27, pp. 483–486.

7. Hui, L., Smith, R., Nelson, J. K., and Schadle, L. S. (2009). Electrochemical Treeing in XLPE/Silica Nanocomposites, *Annual Rept. IEEE CEIDP*, No. 6-2, pp. 111–114.

8. Huang, X., Liu, F., and Jiang, P. (2010). Effect of Nanoparticle Surface Treatment on Morphology, Electrical and Water Treeing Behavior of LLDPE Composites, *IEEE Trans. Dielectr. Electr. Insul.*, **17**(6), pp. 1697–1704.

9. Zou, C., Fothergill, J. C., and Rowe, S. W. (2008). The Effect of Water Absorption on the Dielectric Properties of Epoxy Nanocomposites, *IEEE Trans. Dielectr. Electr. Insul.*, **15**(1), pp. 106–117.

10. Kurimoto, M., Tanaka, T., Murakami, Y., Katayama, T., Yamazaki, T., Murata, Y., Hozumi, Y., and Nagao, M. (2014). Water Tree Retardation in MgO/LDPE Nanocomposite, *IEEJ Trans. Fundamentals Mater.*, **134**(3), pp. 142–147 (in Japanese).

5.8 Insulation Degradation (Material Degradation Due to Tracking)

Takanori Kondo

NGK High Voltage Laboratory, NGK INSULATORS, LTD.,
Komaki-shi, Aichi 485-8566, Japan

t-kondou@ngk.co.jp

Nanocomposites can have enhanced resistance to tracking. Adding nanofillers increases the interfacial region between the nanofillers and polymers and enhances the heat resistance, leading to an improvement in tracking resistance. This property makes nanocomposite materials desirable as outdoor insulators.

5.8.1 Tracking Takes Place by Surface Pollution of Insulators

One of the degradation of an insulator is the tracking degradation. According to IEC 62217 (common test standards for polymer insulators and polymeric hollow-core insulators), tracking is defined as "an irreversible degradation process in which formation of a conducting path starts and develops from the insulator surface" [1].

The mechanism is as follows. When a contamination attaches to the insulator's surface and is moistened, the surface insulation resistance is reduced, and leakage current occurs. The heat created by the leakage current dries the contamination, forming a dry band. The heat by the electric discharge of the dry band gives rise to a tracking on the insulator's surface (carbon track). Tracking degradation is the deterioration of the insulator by formation of tracking. On the other hand, if the heat from the dry band discharge does not form a tracking but erodes the surface, it is called erosion degradation.

The molecular structure of silicone rubber, often used for polymer insulators, has a silicone–oxygen bond as the main chain and makes its carbon ratio lower than that of polymers with carbon-carbon bond as the main chain. This makes silicone rubber more resistant to tracking and erosion. When used near high-contamination areas such as shorelines and industrial districts, however, silicone polymers can have tracking and erosion.

In order to solve this issue, attempts have been made to add a large amount of inorganic fillers, usually of micro size. In general, inorganic fillers have high heat resistance, and tracking and erosion can be reduced by adding such fillers and increasing the heat resistance. In particular, fillers with crystallization water such as ATH (alumina tri hydrate, $Al_2O_3 \cdot 3H_2O$), water in the molecular breaks away when heated. This heat absorption effect reduces the heating of the polymer, reducing tracking and erosion degradations.

On the other hand, recently studies have shown that inorganic nanofillers could reduce tracking and erosion degradation as do inorganic microfillers, or even more effectively.

5.8.2 Inclined Plane Test and Arcing Test Are the Standard Test Methods to Evaluate Tracking and Erosion Resistance of Insulators

As methods for evaluating the resistance to tracking and erosion degradations, there are the inclined plate test (IEC 60587, JIS C2136) and the arc test (IEC 61621, JIS C2135) [2, 3].

In an inclined plane test, a sample of length 120 mm, width 50 mm, and thickness 6 mm is used. Electrodes are connected at the top and bottom of the sample and a filter paper is applied to the top electrode. As shown in Fig. 5.65, the plate is placed at an incline of 45°, a voltage (2.5–4.5 kV) is applied, and contamination solution (ammonium chloride solution with a surfactant) is dropped on the top electrode at certain intervals. As the contamination solution flows on the surface of the sample and reaches the bottom electrode, an electric discharge occurs, inducing tracking and erosion. This process is carried out for 6 h, and the resistance to tracking and erosion degradations is measured by the length of the tracking, the depth of the erosion, and the amount of the leakage current.

In an arc test, two tungsten electrodes are placed at opposing sides on the surface of the insulating material, as shown in Fig. 5.66. A high voltage and low current are applied, and the time it takes for tracking to form is measured. Arcs through the air occur if the insulation of the sample is kept, but they stop once the tracking forms and the insulation breaks down. The time it takes for the arc to stop is measured. This time is defined as the arc resistance arc-proof time, and the arc evaluation is done by this measure. The thickness of the sample is regulated at 3.0 ± 0.25 mm, and the test is over if the erosion penetrates the sample.

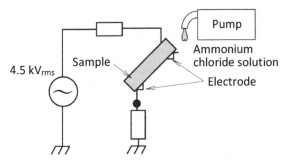

Figure 5.65 Schematic diagram of the inclined plane test.

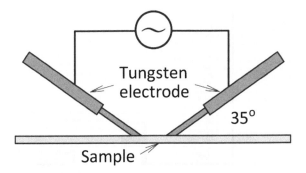

Figure 5.66 Schematic diagram of the arc test.

The above tests evaluate the material's resistance to tracking and erosion. The first method uses contamination solution and thus evaluates the material in a moist condition, while the second method evaluates the material in a dry condition. Outdoor insulators are required to be resistance to tracking and erosion in both wet and dry conditions.

5.8.3 Tracking and Erosion Resistance Is Greatly Enhanced by Nanofiller Addition

There have been studies on improving the resistance to erosion by adding different types of nanofillers to silicone rubber, common used as outdoor insulator, in particular as polymer insulator.

For example, as mentioned in Section 2.3.3, it is known that adding a small amount of silica to RTV (room temperature vulcanizing) silicone rubber significantly increases its resistance to erosion [4].

Other fillers than silica are also known to enhance erosion resistance. For example, 5 wt% of nano-sized MDH (magnesium di hydroxide, $Mg(OH)_2$) filler in RTV silicone rubber increases the erosion resistance more than do micro-sized ATH fillers.

Figure 5.67 shows the erosion weights after inclined plane test for samples without fillers, with micro-sized ATH, and with nano-sized MDH. The sample with micro-sized ATH had lower erosion weight than did the sample without fillers, but the sample with nano-MDH had further reduction in the erosion weight, about 50% of the sample with micro-ATH [5].

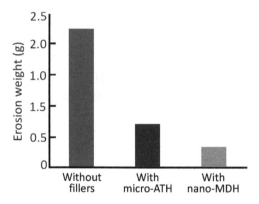

Figure 5.67 Erosion weights after inclined plane test for samples with micro-sized ATH and nano-sized MDH [5].

Improvements in resistance to tracking and erosion have been found in base materials other than silicone polymers. It has been shown that adding nano-sized clay (layered silicate compound) to epoxy resin increases the tracking resistance.

In an inclined plane test with the voltage of 4.5 kV, epoxy resin without fillers had lasted for 1.1 h, and length of the tracking reached the terminating standard of 25 mm. Adding micro-sized silica to epoxy resin with 60 wt% increased the time for tracking to reach 25 mm up to 1.7 h on average.

However, there has been a study in which tracking did not reach 25 mm after the terminating time of 6 h, passing the test, in the material with 50 wt% of micro-sized silica and just 1 wt% of nano-sized clay [6].

The improvements have also been found in arc test. Figure 5.68 shows the relationship between the addition content and the arc proof time for RTV silicone rubber with micro-sized and nano-sized silica. As micro- and nanosilica both increase the arc proof time, the addition amounts increase, improving the arc resistance. At the addition amount of 25 vol%, the sample with nanosilica has the arc proof time twice as long that with microsilica [7].

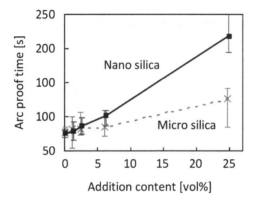

Figure 5.68 Relationship between the addition content and the arc proof time for RTV silicone rubber with micro-sized and nano-sized silica [7].

There have been reports supporting the improvement of tracking and erosion resistance by adding nanofiller to polymers. To summarize the tendency,

(i) the more the filler addition amount,
(ii) the smaller the filler diameter, and
(iii) the higher the filler's dispersiveness,

the more tracking and erosion resistance is enhanced.

Regarding (ii), Figs. 5.69 and 5.70 show the relationship between the filler diameter and the erosion length and the arc proof time, respectively. In the inclined plane test and arc test with 5 wt% silica of diameters 7 nm, 40 nm, 0.5 µm, and 22 µm, the erosion length became shorter and the arc proof time longer as the silica diameter became smaller [8].

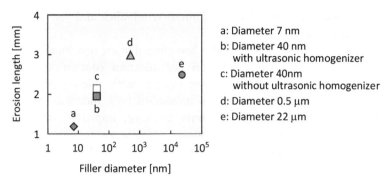

Figure 5.69 Relationship between the filler diameter and the erosion length for samples with 5 wt% in the filler content [8].

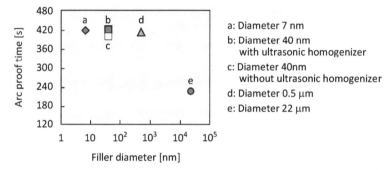

Figure 5.70 Relationship between the filler diameter and the arc proof time for samples with 5 wt% in the filler content [8].

As for (iii), one experiment was done where two samples b and c had 5 wt% silica of diameter 40 nm, but dispersiveness of sample b was improved by an ultrasonic homogenizer in the making process. The improvements in the erosion length and the arc proof time are more prominent in sample b than in sample c, demonstrating that the improvement in the dispersiveness by the ultrasonic homogenizer contributed to enhancing the tracking and erosion resistance. Nanofillers are known to condensate easily, and avoiding condensation is key to nanofiller addition.

5.8.4 Improvement of Heat Resistance Will Lead to That of Tracking and Erosion Resistance

In the previous section, it was shown the key factors in reducing tracking and erosion degradation by adding nanofillers are

(i) high addition amount, (ii) small filler size, and (iii) enhanced dispersiveness. In other words, it is probable that increased interface between the filler and the polymer leads to improved tracking and erosion resistance. We speculated on how an increased interface can lead to enhanced tracking and erosion resistance.

As stated at the beginning, the cause of tracking and erosion degradations is the heat from electric discharge. Thus, it is expected that materials with nanofiller, which have large filler interface, can absorb heat. Also, nanofiller is dispersed near the surface of the polymer (Fig. 5.71 shows the images of microcomposite and nanocomposite material), and the nanofiller thus educes tracking or erosion as they form. In general, inorganic fillers are more heat-resistant than an organic polymer, and the nanofiller can reduce tracking and erosion by educing them.

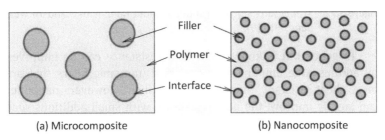

Figure 5.71 Image of microcomposite and nanocomposite material.

As mentioned in the previous chapter, the improvement in heat resistance by adding nanofiller is already known. One of the methods for evaluating the heat resistance is TGA (thermogravimetric analysis). This evaluation method measures the weight change of the sample as its temperature is changed. Materials with high heat resistance have small weight changes. Figure 5.72 shows the TGA results silicone rubber without filler, with 5 and 10 wt% of nano-sized silica.

As the three samples were heated, the weight of the silicone rubber without filler suddenly decreased around 500°. This is due to decomposition of silicone rubber and volatilization of low molecular silicone oil, resulting in weight drop by 65 wt%. However, the weight loss is reduced in the silicone rubber with nanosilica, the sudden drop, as found in the silicone rubber without filler, also did not occur. The decrease was rather smooth. The final weight reduction rates are 28.5 wt% for the sample with 5 wt% nanosilica, 22.5 wt% for that with 10 wt% nanosilica, which are significantly lower than the 65 wt% weight reduction found in the sample without filler [9].

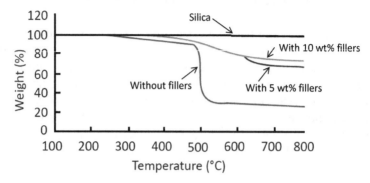

Figure 5.72 TGA results silicone rubber without filler, with 5 and 10 wt% nanosilica [9].

In general, it is known that heat resistance can be improved by adding large amount of micro-sized inorganic fillers that are more heat-resistant than silicone rubber. However, nanofiller can largely improve the heat resistance with small additions such as 5 and 10 wt%. This indicates that the increased interface by adding nanofiller contributes to the enhanced heat resistance.

References

1. The International Electrotechnical Commission. (2012). *IEC-62217*, Polymeric HV Insulators for Indoor and Outdoor Used—General Definitions, Test Methods And Acceptance Criteria.
2. The International Electrotechnical Commission. (2007). *IEC-60587*, Test Methods for Evaluating Resistance to Tracking and Erosion of Electrical Insulating Materials Used Under Severe Ambient Conditions.
3. The International Electrotechnical Commission. (1997). *IEC-61621*, Dry, Solid Insulating Materials. Resistance Test to High-Voltage, Low-Current Arc Discharge.
4. El-Hag, A. H., Jayaram, S. H., and Cherney, E. A. (2004). Comparison between Silicone Rubber containing Micro- and Nano-Size Silica Fillers, *Annual Rept. IEEE CEIDP, No.* 5A-12, pp. 385–388.

5. Venkatesulu, B., and Thomas, M. J. (2008). Studies on the Tracking and Erosion Resistance of RTV Silicone Rubber Nanocomposite, *Annual Rept. IEEE CEIDP,* No. 3A-6, pp. 204–207.

6. Guastavino, F., Thelakkadan, A. S., Coletti, G., and Ratto, A. (2009). Electrical Tracking in Cycloaliphatic Epoxy Based Nanostructured Composites, *Annual Rept. IEEE CEIDP,* No. 7B-16.

7. Raetzke, S., and Kindersberger, J. (2006). The Effect of Interphase Structures in Nanodielectrics, *IEEJ Trans. Fundamentals Mater.,* **126**(11), pp. 1044–1049.

8. Nakamura, T., Kozako, M., Hikita, M., Inoue, R., and Kondo, T. (2012). Effects of Addition of Nano-scale Silica Filler on Erosion, Tracking Resistance and Hydrophobicity of Silicone Rubber, *The 8th International Workshop on High Voltage Engineering,* No. ED-12-143, SP-12-070, HV-12-073, pp. 85–89.

9. El-Hag, A. H., Simon, L. C., Jayaram, S. H., and Cherney, E. A. (2006). Erosion Resistance of Nano-Filled Silicone Rubber, *IEEE Trans. Dielectr. Electr. Insul.,* **13**(1), pp. 122–128.

5.9 Material Degradation Due to Electrochemical Migration

Tsukasa Ohta

Mie University, Tsu-shi, Mie 514-8507, Japan

ohta.tsukasa@mie-u.ac.jp

Electrochemical migration (below shortened to migration) is an insulation deterioration phenomenon for reaching a short circuit by the dissolution as ions and a precipitated product from electrode metals. Recently, since the downsizing and the densification of various electronic parts have advanced, insulation deterioration caused by migration has become more critical. The latest evaluation method verified that nanofiller controls migration.

5.9.1 Why Is a Measure Against the Migration Necessary?

Large computers, OA apparatuses, and much electronic equipment, including both personal and AV products, correspond to lifestyle diversification. These apparatuses continue to be downsized, and their high functions are being simplified. Based on the demand for downsizing and the high efficiency of various electronic parts, high-density implementation of printed wiring boards has been achieved [1].

Figure 5.73 compares the electric field strength of electronic equipment and electric apparatuses. Since thinning or the narrow pitch of printed wiring boards reduced the conductor intervals, and the electric field rose that was applied to a dielectric insulator. The problem of insulation reliability was investigated in a close-up [2]. The fine patterning of printed wiring boards is represented by many stratifications and plane densification [3]. We must focus on wiring board design because the occurring interfaces of migration are greatly increased by many stratifications. Great opportunities for the reliable security are expected for use under the inferior environments of high temperature and high humidity of recent hybrid and battery cars.

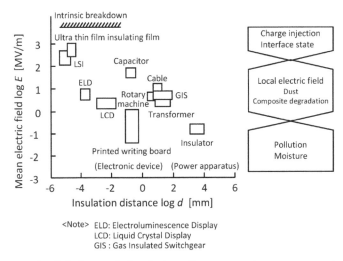

Figure 5.73 Relations of insulation distance and mean use electric field [1].

5.9.2 What Kind of Phenomenon Is the Migration?

Migration is a general term for the shift phenomenon of metal molecules and metal ions. When a DC voltage is applied between two metals of printed wiring boards, the metal ions shift toward the other metal from one metal (an anode) on the surface or on the inside of the insulation board, and the metal or compound becomes a precipitated phenomenon. When the insulation board's surface and inside show an electrolyte property in the presence of water, the metal ions elute it and migration often outbreaks.

The deterioration factors of electronic equipment and electric apparatuses are divided into electric, thermal, mechanical, and electrochemical factors. After adding an environmental condition to these factors, deterioration accelerates. Migration begins with the anode by the electrochemical reaction where the metal is dissolved as an ion. Migration phenomena are classified in Fig. 5.74. In the first case, the metal ions, which left the anode, are reduced and precipitate it, or the compound precipitates it. In addition, in the second case, the metal ions, which were eluted in an anode, reach the cathode where they get an electron and are reduced and deposited.

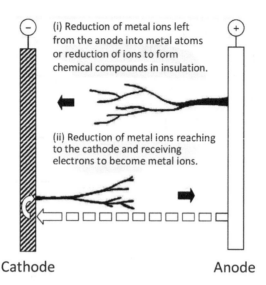

Figure 5.74 Outbreak pattern of migration [4].

When the insulation resistance of the deterioration condition of a high-temperature and high-humidity environment exceeds 10^8 Ω order, the first case is seen. The second case is seen when the insulation resistance decreases more than this. Such a difference in those two case of migration is probably caused by the fact that the degree of the easiness of movement of the metal ions is different. When an ion moves easily, phenomenon of the second case is shown, and phenomenon of the first case seems to be shown when an ion moves with difficulty [4].

One difference of the insulation constitution of printed wiring boards is the conductor wiring. A difference is seen in the migration's outbreak situations between conductors. When the conductor on the board is exposed, migration is seen on the board surface. If there is no coating of the insulating layer, it is easily affected by the external environment regardless of the kind of conductor, and migration patterns are especially different based on the dew condensation situations. An example of migration is shown in Fig. 5.75 when it is coated with an insulating layer.

This migration case is an example of a solder-resist layer and the board interface parts that distinguish whether these distinction uses are reflected light or transmitted light. In such a sample, a Cl element might be detected with Cu in the migration part by analyzing with a scanning electron microscope and an energy dispersive spectroscopy (SEM-EDS) because isolation chlorine accelerates migration. Even if the water that is attached to the board surface is pure, its pH is changed by the kind and the quantity of the material and the gas that is remarkably dissolved from the board might promote migration [5]. Even if migration occurs, we might not adequately understand its condition and the outbreak's cause. Therefore, migration studies advance by making a model that can reasonably explain a problem example of the experiment that occurred with the field.

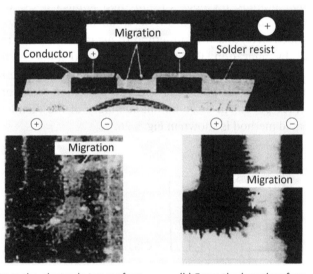

(a) From the electrode top surface   (b) From the board surface

Figure 5.75 Migration that is coated with an insulating layer [4].

5.9.3 Various Reliability Test Methods Are Available for the Migration

Various test methods about migration suggested so far are introduced this section. Originally most of these migration test methods were carried out for the purpose of others. The reliability test method is classified in a simplified test method and an environment testing method roughly. In addition, it is necessary for these test methods to be evaluated in comparison with a real trouble result.

The simplified test method is a method evaluated easy outbreak of migration by the difference of the kind of electrode metal and the insulator in a short time and is effective for the comparison of relative migration characteristic. This method is mainly used as a screening examination of the conductor metal to relatively compare the migration deterioration life in a short time. Because it is really examined in the condition that is considerably more severe than environment condition to be used, this examination has danger to underestimate the materials.

An example of the simplified test method includes the solution dipping method. This method applies a DC voltage in the condition that soaked an electrode in deionized water or a thin electrolyte, and it is a method to measure the time to a short circuit by migration. Besides the method to immerse to a thin electrolyte, there are the distilled water-mediated method and deionized water dripping method and a method to absorb water to a filter paper. An examination summary of the distilled water-mediated method is shown in Fig. 5.76.

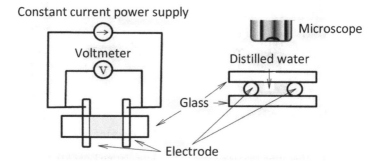

Figure 5.76 Conception diagram of the distilled water-mediated method.

The environment testing method compares the time to a short circuit between conductors by migration when a DC voltage was applied to a printed wiring board sample in a high-temperature and high-humidity atmosphere. In this case, it is carried out under a condition that is more severe than an actual use environment and placed as an acceleration examination to examine the device's life by migration. In this test method, we conducted a steady test, a cycle test, and a dew condensation examination. The pressure cooker test (PCT) has been adopted to evaluate such electronic parts as IC packages. For printed wiring boards, the highly accelerated temperature and humidity stress test (HAST), which is mainly performed and adopted as a severe endurance test, has been adopted not only in the evaluation of electronic parts but also in such evaluations of printed wiring boards.

There are some test standards for methods of environmental tests. The measurement standard substitutes a standard about the insulation resistance measurements. In high-temperature and high-humidity environments, some methods fix the time when a conductor interval short-circuits by continually measuring the leak electric current between conductors. This test is often used as a long-time life evaluation of the samples, and the temperature, the humidity, and the electric field strength were chosen as environmental parameters [6]. The steady test compares the time before a short circuit between conductors in a condition that applied a DC voltage under a constant high-temperature and high-humidity environment.

In addition, the cycle test method changes such atmosphere conditions as temperature and humidity. Because dew condensation might be generated on the sample surface with a temperature change in the cycle test, both the environment tank capacity and the thermal capacity must be examined of the sample enough not to condense dew.

A comb-form electrode is generally used for the conductor patterns of the sample. The shape and the dimensions of the comb-form electrode are listed in IEC technical report 62866, and such electrode parameters as width, interval, and length are decided based on real use conditions. Since the use of mobile type electronic equipment continues to increase, for use in environments that suddenly change, insulation deterioration caused by dew condensation might occur. Based on such a problem, we made dew condensation environments and performed a dew condensation cycle test to evaluate the insulation reliability. Because the migration in dew condensation environments is especially bad in the beginning, it can quickly be evaluated like a simple test method.

5.9.4 The Migration Can Be Evaluated by Measuring the Distribution of Space Charge

The importance of printed wiring boards continues to increase not only in the surface direction of the board but also for the insulation characteristics of the board thickness direction for multilayered high-density wiring structures. From such a view, Ohki et al. detected the following migration that progresses using a pulsed electroacoustic stress (PEA) method in the thickness direction in the insulation composite for printed wiring boards in non-destructiveness [8]. The PEA method applies a pulse electric field to a measured sample, generates a acoustic wave proportional to the space charge, and measures it by detecting the acoustic waves by a piezoelectric element.

The sample's migration examination, which simulated the laminated structure of the printed wiring board in household electrical appliances by the system, is shown in Fig. 5.77. The sample has a compound structure that repeated the pre-pregs which impregnated the phenolic resin to the paper matrix (Fig. 5.78). Epoxy resin (approximately 0.1 mm) was applied as the adhesion layer on a composite, and a copper foil electrode (10 mm × 10 mm, 35 μm thick) was attached to the surface.

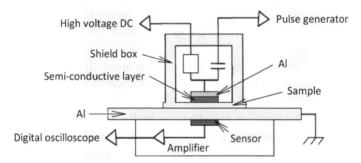

Figure 5.77 PEA measuring system.

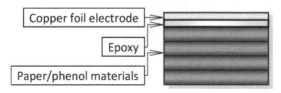

Figure 5.78 Sample.

The examination condition was determined in reference to environment testing standard JPCA-ET04. First, they kept the sample under 85°C, 85%RH for 20 h to preliminarily absorb the moisture. The voltage of a mean 3 kV/mm electric field was applied under the same condition for 100 h. Next they removed a sample from a temperature-humidity test equipment and measured the space charge.

The electric charge distribution that was measured while applying it to the mean electric field 3 kV/mm is shown in Fig. 5.79. The cross axle is the sampling time because the sound velocity of each material is different and cannot uniformly convert thickness. After applying the voltage, a new positive charge peak that seemed to cause migration was observed on the copper anode entire surface. On the other hand, the negative charges decreased that occur in the interface of the epoxy resin layer and the paper/phenol layer.

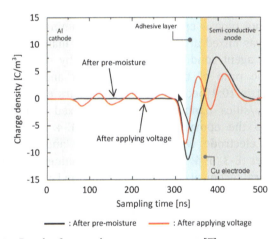

Figure 5.79 Result of space charge measurement [7].

In addition, cross-section SEM-EDS images of the copper foil electrode and an optical microscope image after applying voltage for 100 h are shown in Fig. 5.80. Copper migration advances approximately 50 μm in the thickness direction in Fig. 5.80a and more than 1 mm in the surface direction in Fig. 5.80b. In Fig. 5.80a, copper partly penetrates the epoxy resin layer. Copper migration, which progressed to the epoxy resin layer in this way, probably caused electric charge distribution (Fig. 5.79).

In other words, the conductivity of the epoxy resin layer increased because the difference of the conductivity decreased between the paper/phenolic resin composition and the epoxy resin layers.

(a) Cross-sectional SEM-EDS image (Red dots: Cu)

(b) Optical microscopy image of the sample surface around the anode

Figure 5.80 Cross section SEM and an optical microscope image [7].

5.9.5 Nanocomposites Will Suppress the Migration

Improvement of migration properties was reported by epoxy resin that became a nanocomposite [8, 9]. An experiment sample is shown in Table 5.5. Samples E-3 and E-4 are cured epoxy resin in a bisphenol type A epoxy resin and an amine-type hardening agent. E-3 was mixed at a ratio of 10:3 as a main ingredient and a hardening agent condition recommended by the manufacturer. However, the ratio of the main ingredient and the hardening agent of E-4 is 10:4 because migration occurs easily. Sample NS is an epoxy/silica nanocomposite that mixed a 1 wt% silica nanofiller to the epoxy resin as sample E-4. A 35 μm thick copper foil electrode was attached to the anode side surface with samples E-3, E-4, and NS. They examined the migration under identical technical conditions as explained above.

Figure 5.81 shows the measurement results of the space charge distribution of each sample. The four measurement data (25, 50, 75, and 100 h) are displayed together. As seen in sample E-4 in Fig. 5.81b, the positive charge peaks close to 108 ns and 117 ns decrease as more time was applied to the voltage. Therefore, these electric charges are regarded as the accumulation of cation-related impurities that are included in the epoxy resin from the beginning. In other words, positive ionic impurities move on the cathode, and a positive charge is probably canceled in the negative charge on it. Furthermore, the peak of the positive charge close to 132 ns progresses inside the sample. It seems that since copper is precipitated in the epoxy layer of the anode neighborhood, the conductivity rose. It is believe that this phenomenon caused migration.

Table 5.5 Samples

Sample	Materials
E-3	Epoxy : Hardener = 10 : 3 (106 μm) + Cu (35 μm)
E-4	Epoxy : Hardener = 10 : 4 (88 μm) + Cu (35 μm)
NS	Nanocomposite : Hardener = 10 : 4 (136 μm) + Cu (35 μm)

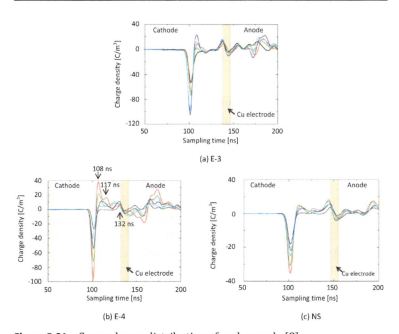

Figure 5.81 Space charge distribution of each sample [8].

As seen in the SEM of Fig. 5.82(b-1), (b-2)and the SEM-EDS image, copper is precipitated from the copper foil layer to the epoxy resin layer, and progresses to the epoxy resin layer by migration. This extension width is approximately 15 μm. In sample E-4 of Fig. 5.81b, the peak of the positive charge moved 6 ns from the 132 ns anode to the 126 ns. The conversion distance of this sampling time was approximately 18 μm and is equivalent to the distance of the SEM image.

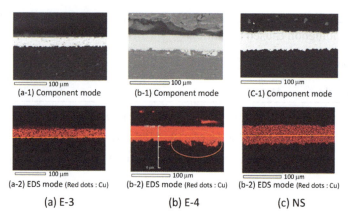

Figure 5.82 SEM-EDS images after the electric field application for 100 h [8].

As seen in sample NS in Fig. 5.81c, the position of the peak of the positive charge on the anode did not move in the sample without changing from the 140 ns position. Similarly, the SEM and SEM-EDS images failed to confirm that copper progressed from the copper foil layer to the epoxy resin layer. The accumulation of the positive charge of 110 ns of Fig. 5.81c corresponds to the positive ionic impurities included in the epoxy resin, such as sample E-4, from the initial condition of epoxy resin. It is confirmed that the migration characteristics of the epoxy resin improved that was added to the silica nanofiller.

A restrain effect of migration is reported not only for epoxy resin but also for nanosilica composite samples of polyimide resin [10]. We showed that migration characteristics are improved by adding nanoparticles, even though the cause is not yet completely elucidated. As more detailed research provides advances, we expect solutions about the function that gives the resin properties of nanofillers.

References

1. Tsukui, T. (1997). Status and Trends of Insulation Reliability Analysis, *J. Japan Inst. Electron. Packaging*, **12**(6), pp. 397–401 (in Japanese).
2. Tsukui, T. (2005). Insulation Deterioration and the Prevention Method by Electrochemical Migration of Electric Equipment (Pt.2), *J. Japan Inst. Electron. Packaging*, **8**(4), pp. 523–530.

3. Tsukui, T. (1999). The Subject of Reliability Evaluation for High Density Electronics Packaging, *IEEJ Trans. Fundamentals Mater.*, **119**(5), pp. 541–546 (in Japanese).

4. Tsukui, T. (2005). Insulation Deterioration and the Prevention Method by Electrochemical Migration of Electric Equipment (Pt.1), *J. Japan Inst. Electron. Packaging*, **8**(4), pp. 339–345.

5. Technical Report of IEC TR 62866. (2014). Electrochemical Migration in Printed Wiring Boards and Assemblies-Mechanisms and Testing.

6. Technical Report of IEE Japan. (1996). No. 615, pp. 9–21.

7. Asakawa, H., Natsui, M., Tanaka, T., Ohki, Y., Maeno, T., and Okamoto, K. (2011). Detection of Electrochemical Migration Grown in a Two-layered Dielectric by the Pulsed Electroacoustic Method and Numerical Analysis of the Signals, *IEEJ Trans. Fundamentals Mater.*, **131**(9), pp. 771–777 (in Japanese).

8. Ohki, Y., Asakawa, H., Wada, G., Yuichi, H., Tanaka, T., Maeno, T., and Okamoto, K. (2011). Several Pieces of Knowledge on Electrochemical Migration Resistance of Printed Wiring Boards Obtained by Space Charge Distribution Measurements, *IEEJ the Paper of Technical Meeting on Dielectrics and Electrical Insulation*, No. DEI-11-079, pp. 13–18 (in Japanese).

9. Ohki, Y., Hirose, Y., Wada, G., Asakawa, H., Tanaka, T., Maeno, T., and Okamoto, K. (2012). Two Methods for Improving Electrochemical Migration Resistance of Printed Wiring Boards, *IEEE International Conference on High Voltage Engineering and Application*, pp. 687–691.

10. Technology and Application of polymer Nanocomposites as Dielectric and Electrical Insulation, *Technical Report of IEE Japan*, No. 1051, p. 34 (in Japanese).

Chapter 6

Thermal and Mechanical Performance of Nanocomposite Insulating Materials

Satoru Hishikawa[a], Shigenori Okada[b], Takahiro Imai[c] and Toshio Shimizu[c]

[a]*Huntsman Japan*
[b]*Takaoka Chemical Co., Ltd.*
[c]*Toshiba Corporation*

6.1 Thermal Performance

Satoru Hishikawa

Application Development, Advanced Materials, Huntsman Japan, Kobe-shi, Hyogo 650-0047, Japan

satoru_hishikawa@huntsman.com

Thermal characteristics, including heat resistance, of insulation materials (polymers) change due to the addition and dispersion of a few wt% of a nanofiller into a polymer. This phenomenon has not been confirmed for the dispersion of a few wt% of the microfiller. The results of the changes in the thermal characteristics

---

Advanced Nanodielectrics: Fundamentals and Applications
Edited by Toshikatsu Tanaka and Takahiro Imai
Copyright © 2017 Pan Stanford Publishing Pte. Ltd.
ISBN 978-981-4745-02-4 (Hardcover), 978-1-315-23074-0 (eBook)
www.panstanford.com

of insulation materials incorporating nanofillers measured by differential scanning calorimetry (DSC) and thermogravimetric analysis (TGA) are reported. The results demonstrate that the glass transition temperature (T_g) and the thermal degradation temperature of polymers can be improved.

6.1.1 Thermal Characteristics Comprise Thermal Behavior, Thermal Properties, and Heat Resistance

Thermal characteristics of polymers, because they can also affect various other characteristics such as electrical and mechanical properties, are the most important material properties describing the nature of the polymers. Thermal characteristics can be broadly classified into thermal behavior (change in state due to temperature change), thermal properties (properties relating to heat including heat transfer and thermal expansion), and heat resistance (properties relating to thermal stability). The typical properties of each characteristic are listed in Table 6.1.

Table 6.1 Typical thermal properties of polymer

Thermal characteristics	Properties
Thermal behavior (Change in state due to temperature change)	Glass transition temperature
	Melting point
	Crystallization temperature
Thermal properties (Properties relating to heat including heat transfer, thermal expansion)	Specific heat
	Thermal conductivity
	Coefficient of thermal expansion
Heat resistance (Properties relating to heat stability)	Thermal degradation temperature
	Weight loss on heating
	Deflection temperature under load
	Continuous service temperature

As an example, thermosetting resins such as epoxy resin have a glassy nature due to the limited motions of molecular chains at temperatures below a certain value, but the resins become soft and rubbery due to the free motion of the molecular chains above the certain temperature as a result of heating. This temperature is called the glass transition temperature. Thermal degradation temperature is the temperature at which the polymer begins to decompose by heating while it is gradually heated. Both temperatures are important indexes in the selection of an insulation material for use in electrical apparatus and electronic devices, and for designing insulation systems.

As the performances of insulating materials have been improved, it has become obvious that polymer thermal characteristics change as a result of the inclusion of a few wt% of the nanofiller in polymers. The effects of nanofiller dispersion on the thermal characteristics of the polymer, including on the thermal behavior and heat resistance, are described below.

6.1.2 Epoxy Resin Thermal Characteristics Can Be Altered Using a Nanofiller

Many studies relating to changes in the thermal characteristics of polymer nanocomposites have previously been reported. In this section, the T_g changes of nanocomposites based on epoxy resin are introduced.

6.1.2.1 Effects of several types of nanofillers on the glass transition temperature

The changes in the glass transition temperature (measured by DSC) of nanocomposites by the dispersion of several types of nanofillers, including TiO_2, Al_2O_3, and ZnO (0.1–10 wt%), into the matrix of Bisphenol A type epoxy resin/aliphatic amine curing agent has been reported [1]. Table 6.2 shows the details of test samples and Fig. 6.1 shows the T_g results for each test sample.

From the figure, the dispersion of TiO_2 microfiller (1 wt% and 5 wt%) into epoxy resin does not cause any change in the T_g values compared with that in the base matrix. In contrast, in nanocomposites for all types of fillers, there is a gradual decrease in T_g values up to 0.5 wt% filler concentration. Beyond 0.5 wt% filler loading, T_g tends to increase up to 5 wt% filler loading with the exception of Al_2O_3 nanofillers (5 wt% filler loading).

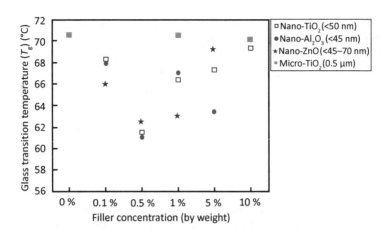

Figure 6.1 Variations of glass transition temperatures in epoxy nanocomposites with respect to filler concentration [1].

Table 6.2 Epoxy nanocomposites sample by dispersion of nanofillers

Epoxy resin (CY1300)/aliphatic amine (HY956) = 100: 25 (parts by weight)		
Kinds of filler		**Average particle size**
Nanofiller	TiO_2	ca. 50 nm
	Al_2O_3	ca. 45 nm
	ZnO	ca. 45–70 nm
Microfiller	TiO_2	ca. 0.5 μm

Nanocomposites sample preparation:
(1) Degassing of the materials to remove moisture (Resin, Hardener: 40°C/2 h under degassing, Filler: 90°C/24 h)
(2) Mixing of resin/nanofiller with a high shear mechanical mixer (700 rpm/1 h)
(3) After degassing, mixing with Ultrasonic technique (24 kHz) and then mixing hardener with hand
(4) Casting the mixture into the mold and then degassing → curing at 60°C/4 h (For microfiller, mixing with a high shear mechanical mixer at 700 rpm/120 s)

6.1.2.2 Effects of several types of nanofillers on the glass transition temperature

The changes in T_g of nanocomposites by inclusion of a nanofiller (0–20 wt%, average diameter: 25 nm) in the matrix of Bisphenol

A type epoxy resin/anhydride curing agent with flexibilizer (20 wt%) is reported [2].

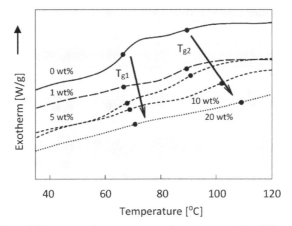

Figure 6.2 DSC curves of pure epoxy and nanocomposites [2].

Figure 6.2 illustrates the DSC curve of pure epoxy and nanocomposites and Table 6.3 lists the T_g value for pure epoxy and nanocomposites. A large amount of flexibilizer (20%) is added to the base matrix to increase its flexibility; thus, the cured material comprises a tightly cross-linked network and loose polymer chains with many small molecules (flexibilizer) dispersed in the chains. Two glass transition temperatures are present on each DSC curve (Fig. 6.2). T_{g2} is associated with the glass transition of the high-density cross-link network, and T_{g1} represents the chain movement in loose structures consisting of free epoxy chains. From Table 6.3, the values of T_{g1} and T_{g2} increase with the nanofiller concentration. From Fig. 6.2, the transition at T_{g1} is weakened in the nanocomposites, which reflects the fact that the flexibility of the nanocomposite reduces and the degree of cross-linking increases.

From the analysis of T_g by DSC and other methods, the structures of the cross-linked network in cured epoxy resin with and without silica nanofillers can be obtained, as illustrated in Fig. 6.3. The dispersed nanofillers fill the free volume, that is formed by the addition of flexibilizer, and increase the degree of cross-linking; thus, it is considered that T_g increases with the concentration of the silica nanofiller.

Table 6.3 Glass transition temperatures in pure epoxy and nanocomposites [2]

Silica nanofiller concentration [wt%]	0	1	5	10	20
T_{g1} [°C]	66.4	66.9	68.2	69.3	71.6
T_{g2} [°C]	89.6	89.6	90.4	101.8	108.8

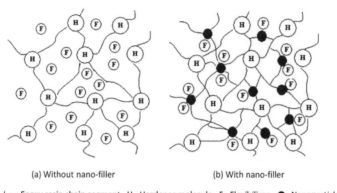

(a) Without nano-filler (b) With nano-filler

(∼: Epoxy resin chain segment, H : Hardener molecule, F : Flexibilizer, ●: Nanoparticle)

Figure 6.3 Schematic of cross-linking structure in cured epoxy nanocomposites [2].

6.1.2.3 Effects of nanofiller dispersion on the glass transition temperature of nanomicrocomposites

Changes in the glass transition temperature of nanocomposites (N) by dispersion of nanoclay and of nanomicrocomposites (NM) by dispersion of nanoclay and microsilica into the matrix of bisphenol A type epoxy resin/anhydride (MHHPA) are investigated.

Table 6.4 lists the sample names and filler details; Fig. 6.4 shows the DSC spectra for each sample. The initiation temperature (T_{ig}), the intermediate temperature (T_{mg}), and the end temperature (T_{eg}) in nanocomposites (N1) with a clay nanofiller organically modified by primary amine are lower than the equivalent temperatures in the base epoxy resin (E). Compared with silica microcomposites (M), T_{ig}, T_{mg}, and T_{eg} are lower in nanomicrocomposites (NM1) with the clay nanofiller and silica microfiller. In contrast, T_{ig} and T_{mg} in nanomicrocomposites (NM3) with the clay nanofiller organically modified by tertiary amine and silica microfiller have similar values to M, but T_{eg} in NM3 is higher. It is considered that the clay nanofiller with organic modification can affect the cross-linking density of epoxy resin, thus causing changes in T_g. Furthermore, the repressed motion of epoxy molecular chains caused by the higher content of silica microfiller can affect the changes in the T_g value.

Table 6.4 Fillers used in nanocomposites and nanomicrocomposites [3]

Sample code	Filler	Filler content	Nanoclay modifiers	Swelling agent for nanoclay	Silane coupling treatment for microsilica
E	W/O	—	—	—	—
N1	Nanoclay	3.0 vol%	Octadecylamine (primary amine)	Done	—
N3	Nanoclay		Dimethyldodecylamine (Tertiary amine)	Done	—
M	Micro-silica	48.3 vol%	—	—	Done
NM1	Nanoclay Micro-silica	Clay: 1.6 vol% Silica: 47.5 vol%	Octadecylamine (primary amine)	Done	Done
NM3	Nanoclay Micro-silica		Dimethyldodecylamine (Tertiary amine)	Done	Done

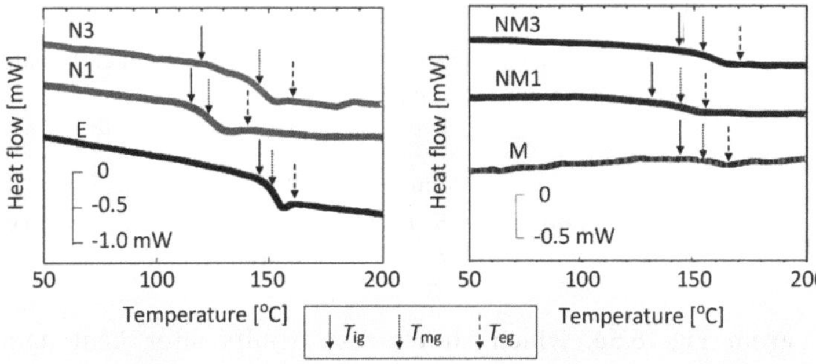

(a) nanocomposites   (b) nano-microcomposites

Figure 6.4 DSC spectra of nanocomposites and nanomicrocomposites [3].

6.1.3 Thermal Characteristics of Several Types of Polymers Change by the Dispersion of the Nanofillers

In the previous section, the effects on T_g of nanofiller dispersion in the polymer as a thermal characteristic of epoxy nanocomposites were introduced. Many studies of changes in the thermal characteristics of several types of polymers that are not epoxy resin by dispersion of the nanofiller have been conducted. The results of some of these studies are described below.

6.1.3.1 Room Temperature Vulcanizing of RTV silicones

Alterations in the thermal characteristics of silicone nanocomposites by dispersion of a silica nanofiller (nano-fumed silica) as a reinforcing material and silicone nanomicrocomposites by the covariance of nano-fumed silica and microfiller (calcium carbonate) as a bulk filler have been described [4]. The details of dispersed fillers are listed in Table 6.5.

Table 6.5 Fillers used in this experiment [4]

Filler	Trade name	Particle size and specific surface area	Remarks
Reinforcing	R972 (Nippon Aerosil Co., Ltd.)	16 nm 130 m²/g	Surface treatment by dimethylsilane
Extender	Super #1700 (Maruo Calcium Co., Ltd.)	1.3 µm 1.7 m²/g	Ground calcium carbonate
	Viscolite-OS (Shiraishi Kogyo Kaisha, Ltd.)	80 nm 15.5–18.5 m²/g	Precipitated calcium carbonate and the surface is treated by fatty acid

From Fig. 6.5a, which shows test results after heat aging at 100°C, silicone nanomicrocomposites with further loading of calcium carbonate (#1700) in addition to nano-fumed silica (R972), compared with nanocomposites with nano-fumed silica, show an increase in residual weight after being incubated at

100°C for 20 days. Nanomicrocomposites with further loading of calcium carbonate (#1700) in addition to nano-fumed silica (R972) did not show changes in weight loss; however, nanomicrocomposites with additional calcium carbonate (Viscolite) displayed a large weight loss at 200°C for over 150 h as illustrated in Fig. 6.5b, which depicts the test results after heat aging at 200°C. From TGA, oxidative degradation of the fatty acids used as a surface modifier for calcium carbonate (Viscolite) was investigated from around 250°C. Decomposition of fatty acids may generate free radicals, which can lead to the degradation of silicones by peroxidation.

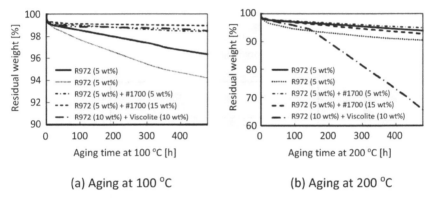

(a) Aging at 100 °C (b) Aging at 200 °C

Figure 6.5 Residual weight of silicone composites at elevated temperature for 20 days [4].

6.1.3.2 Polypropylene

Thermal endurance characteristics of isotactic polypropylene (iPP) nanocomposites incorporating a layered silicate (LS) have been described [5]. Figure 6.6 shows the test results for thermal endurance obtained from isothermal thermogravimetric analysis (ITA).

The time corresponding to a 5% weight decrease in the 130–150°C range for iPP/LS nanocomposites is markedly longer compared with the value for base iPP (ca. 5 times at 130°C: ca. 10 times at 150°C). Figure 6.7 shows the results of the time corresponding to 5% weight loss as a function of test temperature. The temperature index (IT), which is the point of 5% weight loss after

20,000 h according to IEC 60126, was calculated. IT of iPP/LS is ca. 8°C higher than that of the base iPP; hence, the designed temperature of an insulation system using nanostructured iPP can be upgraded to 8°C higher than that of the base iPP.

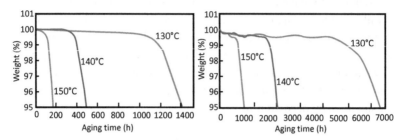

Figure 6.6 Thermal endurance properties (ITA) for base iPP and nanostructured iPP [5]. (a) Base isotactic polypropylene. (b) Isotactic polypropylene/layered silicate.

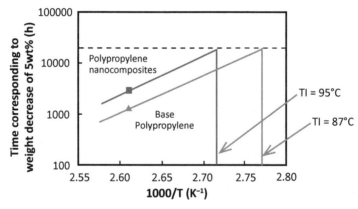

Figure 6.7 Time corresponding to weight decrease of 5 wt% as a function of aging temperature (Arrhenius plots) [5].

6.1.3.3 Polyethylene

Changes in the thermal properties of nanocomposites resulting from dispersion of the surface-treated nanofiller (SiO_2) have been described in a previous study [6]. Figure 6.8 shows the TGA results of polyethylene (PE)/SiO_2 nanocomposites. PE is the base polymer, PE/SiO_2-1 is a nanocomposite containing the surface-untreated silica nanofiller, and PE/SiO_2-2 to PE/SiO_2-4 are nanocomposites with various surface-treated silica nanofillers.

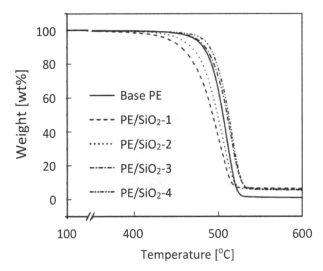

Figure 6.8 TG curves of PE nanocomposites. © 2006 IEEE. Reprinted, with permission, from Han, Z., Diao, C., Li, Y. and Zhao, H. (2006). Thermal Properties of LDPE/Silica Nanocomposites, *Annual Rept. IEEE CEIDP*, No. 3B-5, pp. 310–312.

By incorporating a silica nanofiller into pure PE, the temperature at which the weight reduction occurs (i.e., the thermal degradation temperature) changes. The thermal degradation temperature of PE/SiO$_2$-1 with a dispersed untreated silica nanofiller is lower than that of pure PE. On the other hand, the dispersibility of the silica nanofiller can be improved by appropriate surface treatment, and hence, the thermal degradation temperatures of PE/SiO$_2$-3 and PE/SiO$_2$-4 are higher than that of pure PE.

6.1.4 Polymer Thermal Characteristics Can Be Changed by Interfacial Interactions between Nanofillers and Polymer

Attractive, repulsive, and neutral interfacial reactions can take place between nanofillers and polymers. The thermal characteristics of polymers can be altered by these interactions.

Figure 6.9 shows the theoretical interfacial interactions between a typical polymer (epoxy resin) and nanofiller. The interactions between the polymer and nanofiller lead to the formation of two nanolayers around the nanofiller. The first nanolayer, which is closest to the nanofiller surface (innermost nanolayer), is assumed to be tightly bound to the surface, resulting in the polymer chains there being highly immobile. The second polymer nanolayer is slightly thicker than the first layer and contains polymer chains that are loosely bound.

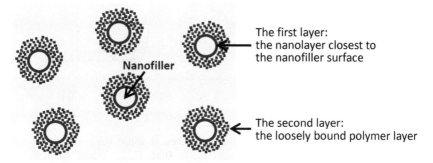

Figure 6.9 Dual-layer model in nanocomposites [1].

The loosely bound polymer in the extended layer causes a reduction in the thermal characteristics of the nanocomposite, such as the glass transition temperature, at low nanofiller concentrations (i.e., with low nanofiller dispersion). As the filler concentration increases (i.e., as the amount of nanofiller dispersion increases), the inter-particle distances begin to decrease, which can also cause overlap of the immobile polymer regions around the nanofillers. This increase in the volume of overlapping immobile polymer regions in the nanocomposites will increases the glass transition temperature and the thermal degradation temperature; thus, the thermal characteristics of the polymer will be improved. Therefore, the interfacial interactions between the polymer chains and the nanofiller lead to the formation of two layers around the nanofiller, and the outer loosely bound polymer nanolayer causes the thermal characteristics to change.

As mentioned above, the effects of nanofiller dispersion on polymer thermal characteristics are introduced. It is very interesting that the thermal characteristics, which are among the most important material properties, can be changed by the addition of a nanofiller. As the thermal characteristics of polymers can be improved, or, conversely, reduced, using nanofillers, it is necessary to consider the type, surface treatments, and amount of nanofiller and also the polymer type (and the affinity of the nanofiller for the polymer) when selecting a polymer for a particular purpose.

References

1. Singha, S. and Thomas, M. J. (2008). Dielectric Properties of Epoxy Nanocomposites, *IEEE Trans. Dielectr. Electr. Insul.*, **15**, No. 1, pp. 12–23.

2. Xu, M., Montanari, G. C., Fabiani, D., Dissado, L. A. and Krivda, A. (2011). Supporting the Electromechanical Nature of Ultra-Fast Charge Pulses in Insulating Polymer Conduction, *Proc. IEEJ ISEIM*, No. A-1, pp. 1–4.

3. Hyuga, M., Tanaka, T., Ohki, Y., Imai, T., Harada, M., Ochi, M. (2011). Correlation between Mechanical and Dielectric Relaxation Processes in Epoxy Resin Composites with Nano- and Micro-Fillers, *IEEJ Trans. Fundamentals and Materials*, **131**, No. 12, pp. 1041–1047 (In Japanese).

4. Cho, H., Ashida, Y., Nakamura, S. and Murakami, Y. (2011). Improvement of Heat-resistance of RTV Silicone Elastomers with Reduce Environmental Impact by Loading Nano-silica and Calcium Carbonate, *Proc. IEEJ ISEIM*, No. MVP 2-5, pp. 345–348.

5. Motori, A., Patuelli, F., Saccani, A., Montanari, G. C. and Mulhaupt, R. (2005). Improving Thermal Endurance Properties of Polypropylene by Nanostructuration, *Annual Rept. IEEE CEIDP*, No. 2C-13, pp. 195–198.

6. Han, Z., Diao, C., Li, Y. and Zhao, H. (2006). Thermal Properties of LDPE/silica Nanocomposites, *Annual Rept. IEEE CEIDP*, No. 3B-5, pp. 310–312.

6.2 Mechanical Performance

Shigenori Okada[a] and Takahiro Imai[b]

[a]*Takaoka Chemical Co., Ltd.*
Ama-shi, Aichi 490-1111, Japan
[b]*Power and Industrial Systems R&D Center, Toshiba Corporation,*
Fuchu-shi, Tokyo 183-8511, Japan

okada.shigenori@tktk.co.jp, takahiro2.imai@toshiba.co.jp

Polymers (plastics) are light and easy to mold compared with metals; however, they often fall short with regard to mechanical characteristics. To improve these mechanical characteristics (which are insufficient in a single polymer), polymers are often combined with one or more materials as a composite material. Composites with nanoscale components are called polymer nanocomposites. For polyamide (nylon) nanocomposites—the first polymer nanocomposites developed—the tensile strength was reported to have improved. Following this discovery, the improvements in mechanical characteristics such as bending strength and fracture toughness in addition to tensile strength have been confirmed to occur when materials are converted into nanocomposites.

6.2.1 Polymer Composites with Improved Mechanical Performance Are Utilized in Daily Life

The idea of improving mechanical characteristics of a single polymer (plastic) by combining it with one or more materials to create a composite material has been around for a long time. Examples include mud walls supplemented by clay and straw, and composite bows made using wood or bamboo strengthened by bones, tendons, and glue. More recent and familiar examples of products with improved mechanical characteristics are vulcanized rubber compound with carbon black, tires reinforced with fibers, unit bathtubs, tennis rackets, and golf club shafts.

Representative modern composite materials include glass fiber–reinforced plastics (GFRPs) and carbon fiber–reinforced plastics (CFRPs), which have been developed using epoxy resins. Unit bathtubs are made using GFRPs, while tennis rackets and golf club shafts are made with CFRPs. CFRPs are particularly attracting attention as light and strong materials with high elastic moduli for use in aircrafts.

Today, the idea of composite materials has widely spread and includes materials filled with fine fillers and polymer alloy materials (made using multiple types of polymers). Table 6.6 shows the sizes of reinforcing materials in composite materials such as fillers and fibers vary [1]. For example, calcium carbonate, used in rubber reinforcement, is often applied using particles with diameters larger than 1 μm, and fibers used in GFRPs and CFRPs have diameters of approximately 10 μm.

Table 6.6   Comparison of reinforcement size in polymer composites

Composite		Reinforcement	Size of reinforcement				
			nm			μm	
			1	10	100	1	10
Reinforced rubber		Calcium carbonate				■	■
		Carbon black			■		
		Glass fiber					■
Reinforced plastic	Impact-resistant polymer alloy material (ABS resin)	Polybutadiene			■		
	Polymer alloy (polymer blend) material	Additional polymer				■	
	Fiber-reinforced plastic	Glass fiber, Carbon fiber, etc.				■	■
	Epoxy-based cast resin for power application	Silica, Alumina, etc.					■
	Epoxy-based encapsulation resin for electronic device	Silica, Alumina, etc.					■
	Polyamide/clay nanocomposite	Clay		■	■		

On the one hand, the diameter of polybutadiene latex used in ABS resin, which is one of the most common shock-resistant polymer alloy materials, ranges from 100 nm to 1 μm. Silica and alumina fillers, used as a casting resin for electrical equipment and as a sealing resin for semiconductors, have diameters of a few tens of micrometers. The size of clay used in polyamide/clay nanocomposites ranges from a few nanometers to a few hundred nanometers: these particles are quite small.

For FRPs, the mass ratio of fiber to polymer is 20–60%, whereas for electrical equipment casting resin and semiconductor sealing resin, the ratios of silica and alumina fillers are 60–90%, indicating higher reinforcement content. In contrast, for polyamide/clay nanocomposites, the mass ratio of the reinforcement material (clay) is less than 10%, indicating a better efficiency for improving the mechanical characteristics of polymers.

6.2.2 Various Mechanical Properties Are Classified by Stressing Time Conditions

The mechanical properties of polymers are improved by creating composites; however, there are various characteristics that need to be considered. When an external force is applied to a material, a change in its condition occurs, such as deformation (stretch or bend) or fracture.

As shown in Fig. 6.10, these changes depend on the type, size, and application of the external force. Thus, their characteristic values are numerical values measured by testing methods defined by the Japanese Industrial Standards (JIS) or the American Society of Testing Materials (ASTM) standards.

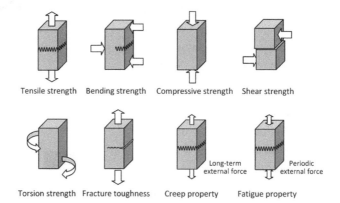

Figure 6.10 Typical mechanical properties of polymers.

In particular, as shown in Table 6.7, mechanical characteristics can be categorized according to parameters such as the speed of external force and the duration of applied force [2].

Table 6.7 Typical mechanical properties of polymers

Stressing time conditions of external force	Mechanical properties
Force applied slowly at a constant speed (Static mechanical properties)	Tensile strength
	Bending strength
	Compressive strength
	Torsion strength
	Elastic modulus (Bending elastic modulus, Young's modulus, etc.)
	Hardness (Rockwell hardness, etc.)
	Fracture toughness
Force applied as an impact (Dynamic mechanical properties)	Impact strength (Charpy impact strength, Izod impact strength, etc.)
Constant force applied for a long duration (Durability of materials)	Creep property (Tensile creep property, Bending creep property, Stress relaxation characteristic etc.)
Force applied periodically (Durability of materials)	Fatigue property (Tensile fatigue property, Bending fatigue property)

6.2.2.1 Force applied slowly at a constant speed

Static mechanical properties of a material can be measured by applying a force at a constant low speed. As shown in Fig. 6.10, common characteristics such as tensile, bending, and compressive strengths can be obtained. In a tension test, for example, the tensile strength is the maximum tensile stress recorded, which is the measured load divided by the cross-sectional area of the material.

6.2.2.2 Force applied as an impact

Dynamic mechanical properties of a material can be obtained by applying a force as an impact (in a significantly short time period). Impact strength is the tolerance to external impacts, and typically defined as the Charpy impact strength or the Izod impact strength for polymers (plastics).

6.2.2.3 Constant force applied for a long duration

The durability of a material can be measured by applying a constant force for an extended period. When a stress lower than the maximum strength (such as that at the fracture point), which can be measured in test [1], is applied, the material slowly deforms with time (the creep phenomenon) and fractures when it reaches the maximum extent of deformation. To test this, a constant force is applied to the material, and the deformation rate (strain) and the fracture point are measured at different temperatures and magnitudes of force from which the durability and the creep lifetime can be evaluated.

6.2.2.4 Force applied periodically

The durability of a material can also be measured by applying a force at periodic intervals. When a stress lower than the maximum strength is applied periodically, the material deteriorates (fatigue) and eventually fractures. To test this, a certain average external force is applied periodically to measure the number of cycles before fracturing (fatigue lifetime). The durability is evaluated by plotting the magnitude of applied force against the number of cycles (S–N plot).

Among the introduced categories, the static characteristics of polymers have been improved by creating nanocomposites from polymers. We now introduce examples of polymer nanocomposites with improved tensile strength, bending strength, fracture toughness, and hardness.

6.2.3 Tensile Strength Increases by Nanofiller Addition

The resistance of a material to a pulling force is called the "tensile strength" and is one of the main representative mechanical characteristics. There have been a number of reports indicating improved tensile strengths with nanofiller dispersion.

For example, Fig. 6.11a shows the effects of the carbon number of alkylammonium ions, used in the organification processes of layered filler (clay), on tensile strength (a-1) and tensile modulus (a-2) of epoxy resin nanocomposites. Materials were formed using epoxy resin hardened by polyether diamine after dispersing organified clay (10 wt%) with $CH_3(CH_2)_7NH_3^+$

(carbon number 8), $CH_3(CH_2)_{11}NH_3^+$ (carbon number 12), or $CH_3(CH_2)_{17}NH_3^+$ (carbon number 17).

The tensile strength and tensile modulus were significantly improved when a carbon chain length in the alkylammonium ions is higher than 12. An increased carbon number of the organification agent leads to enhanced clay's affinity to epoxy resin and thus to a uniform dispersion, improving the tensile characteristics of the nanocomposites.

Figure 6.11b shows the effects of the content of clay that has been through an organification process on tensile strength (b-1) and tensile modulus (b-2) of epoxy resin nanocomposites. As the content of clay, which is organified with $CH_3(CH_2)_{17}NH_3^+$, increases, the tensile strength and tensile modulus also increase. For the case using a clay content of 23 wt%, the tensile strength is 20 times higher and tensile modulus 12 times higher compared with those of the base epoxy resin (zero clay content), indicating that the increased clay content leads to higher reinforcement effect and improvements in tensile characteristics of the nanocomposite.

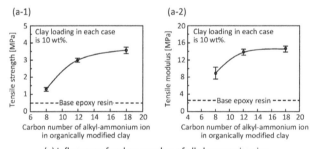

(a) Influences of carbon number of alkyl-ammonium ion in organically modified clay on tensile properties

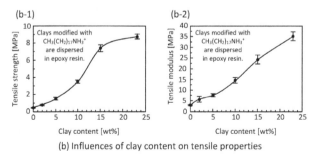

(b) Influences of clay content on tensile properties

Figure 6.11 Tensile properties of epoxy/clay nanocomposites [3].

In general, tensile characteristics of nanocomposites depend on the content, size, and shape of the nanofiller, as well as the resin/filler interface affinity. We now introduce examples of improved tensile properties (Young's modulus) by improving the affinity between resin and nanofiller.

There has been a report investigating Young's moduli of nanocomposites made using ethylene vinyl acetate copolymer (EVA) resin dispersed with barium titanate (BaTiO$_3$) filler, which has a high dielectric constant [4]. EVA resin is both flexible and elastic, and with a dispersion of the BaTiO$_3$ filler, it gains a high dielectric constant, and can be used as an electric field control material in cables and actuator materials.

As shown in Fig. 6.12a, the surface of the BaTiO$_3$ nanofiller, with an average particle diameter of 100 nm, was modified so that hydrogen bonds could form with the EVA resin. The EVA resin nanocomposite was prepared using surface-modified BaTiO$_3$ nanofiller dispersed with a 0–50% volume ratio.

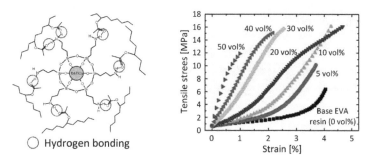

(a) Surface modified BaTiO$_3$ nano-filler (100 nm diameter)

(b) Stress-strain curve of EVA/BaTiO$_3$ nanocomposites

Figure 6.12 Tensile property of EVA/BaTiO$_3$ nanocomposites. © 2010 IEEE. Reprinted, with permission, from Huang, X., Xie, L., Jiang, P., and Liu, F. (2010). Enhancing the Permittivity, Thermal Conductivity and Mechanical Strength of Elastomer Composites by Using Surface Modified BaTiO$_3$ Nanoparticles, *Proc. IEEE ICSD*, No. G3-2, pp. 830–833.

Figure 6.12b shows the resulting stress–strain (S–S) curve. The slope of the linear part of the S–S curve is a mechanical characteristic called Young's modulus (modulus of longitudinal elasticity) and indicates the amount of stress required to form a unit strain in the elastic region of the material. As the nanofiller content in the EVA resin/BaTiO$_3$ nanocomposites increases to 5, 10, 20, 30, 40, and 50 vol%, Young's moduli increases to 1.03, 1.77, 2.33, 4.89, 9.34 and 21.33 MPa, respectively, indicating that it is a material that does not strain easily.

6.2.4 Flexure Performance Improves by Nanofiller Addition

There have been reports on improving a representative mechanical characteristic, the "bending strength," by nanofiller dispersion.

As shown in Fig. 6.13a,b, nanocomposites formed using epoxy resin dispersed with boehmite alumina nanofiller have enhanced bending strength and modulus [5]. When comparing base epoxy resin (0 wt%) with an epoxy composite containing 20 wt% reactive diluent (plasticizer) with 3 wt% nanocomposite, the bending strength and modulus were improved by 5% and 7%, respectively.

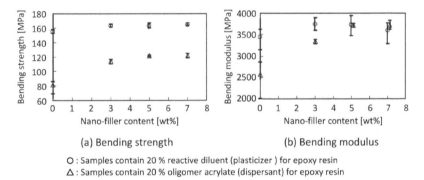

(a) Bending strength   (b) Bending modulus

O : Samples contain 20 % reactive diluent (plasticizer) for epoxy resin
△ : Samples contain 20 % oligomer acrylate (dispersant) for epoxy resin

Figure 6.13 Bending properties of epoxy/boehmite alumina nanocomposites [5].

Using approximately 20 wt% acrylic oligomer (dispersant), and the nanocomposite containing 3 wt% nanofillers has 45% and 40% higher bending strength and modulus than those of the base epoxy resin, respectively. Increasing the boehmite nanofiller content to 5% or 7% showed no significant effect on the bending strength or modulus.

Nanocomposites dispersed with different nanofillers have also been reported to have an increased bending strength.

Figure 6.14a shows the bending strength of a nanocomposite made using epoxy resin with a layered dispersion of clay [6]. The bending strength increased with increasing clay content. Figure 6.14b shows the bending strengths of epoxy resins filled with clay, silica, and titanium dioxide as nanofillers. Using a nanocomposite containing 5 wt% nanofillers, the clay filler showed the highest bending strength.

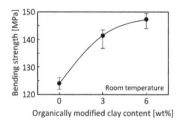 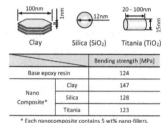

(a) Bending strength of epoxy/clay nanocomposites

(b) Bending strength of nanocomposites containing various kinds of nano-fillers.

Figure 6.14 Influences of nanofillers on bending strength of epoxy based nanocomposites [6].

In general the bending strength of nanocomposites depends on the size and shape of the nanofiller as well as the resin/filler interface affinity. For the same filler contents, clay is larger in size compared with other nanofillers, and has a shape with a higher aspect ratio (of major to minor axes of fillers), leading to a larger reinforcement effect.

6.2.5 Nanofillers Will Suppress Propagation of Crack

One of the static mechanical characteristics is "toughness," which is the resistance to crack propagation, and becomes impact strength when the external force is applied as an impact.

One of the indices for the resistance to crack propagation is fracture toughness (K_{IC}), and as shown in Fig. 6.15, is evaluated using a compact tension (CT) specimen test or a single-edge notched bending (SENB) test. K_{IC} is measured by making an initial crack in the specimen and measuring the load at the fracture point. High K_{IC} values indicate high resistance to crack propagation.

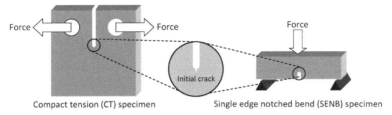

Figure 6.15 Specimens for fracture toughness measurements.

As shown in Fig. 6.16a, reports have revealed improvements in the fracture toughness of polymers dispersed with nanofillers [7]. In a nanocomposite (with silane treatment) made using epoxy resin with 5 wt% silica dispersion of particle size 20–80 nm (dispersed as an aggregate because the primary particle diameter is 12 nm), the fracture toughness was higher than the base epoxy resin by 59% in a CT test. Similarly, a microcomposite (with silane treatment) made using epoxy resin with 5 wt% dispersion of silica microfiller (primary particle diameter 1.6 µm) showed a fracture toughness that was 20% higher than that of the base epoxy resin, implying that the silica nanofiller has a stronger effect on fracture toughness than the silica microfiller.

Moreover, it has been reported that the fracture toughness values of nanocomposites and microcomposites would be decreased if the silica filler surface is not treated with silane coupling agent. Figure 6.16b shows scanning electron microscopy (SEM) images of fracture surfaces near the initial crack of a base epoxy resin, a nanocomposite (with silane treatment), and a microcomposite (with silane treatment). The fracture surfaces are clearly different. The nanocomposite (with silane treatment) showed rough areas in a linear fashion along the direction of crack propagation, while the microcomposite (with silane treatment) showed scattered rough areas. Since the traces (rough areas) of cracks blocked by fillers had linear patterns, it suggested that in nanocomposites (with silane treatment), crack propagation is blocked continuously, which elongates the development distance, and leads to a higher fracture toughness.

Moreover, nanocomposites with a dispersion of layered filler instead of silica nanofiller have also been reported to have increased fracture toughness [6, 8]. As shown in Fig. 6.17, fracture toughness increased with an increasing content of layered filler.

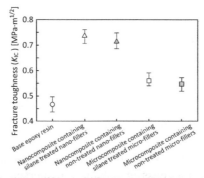

(a) Comparison of fracture toughness in epoxy-based composites

(b) Scanning electron microscopy (SEM) images of fractured surface

Figure 6.16 Fracture toughness of epoxy/silica nanocomposites [7].

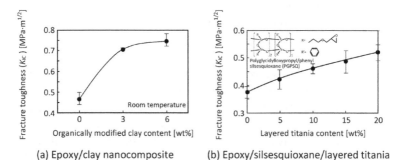

(a) Epoxy/clay nanocomposite

(b) Epoxy/silsesquioxane/layered titania nanocomposites

Figure 6.17 Fracture toughness of epoxy-based nanocomposites containing layered fillers [6, 8].

6.2.6 Various Mechanical Properties Are Reported for Dielectric Nanocomposites

We have introduced tensile strength, bending strength, and fracture toughness as examples of representative characteristics of polymer that are improved as nanocomposites. There are some other mechanical characteristics reported, and we now briefly introduce some of them (see Section 6.3 for long-term durability).

6.2.6.1 Scratch hardness of polyamide nanocomposites

There has been a report on scratch hardness in polyamide nanocomposite films with the dispersion of a 40 nm nanofiller [9]. Scratch hardness is measured by making a scratch of certain width and measuring the load required to make the scratch. As shown in Fig. 6.18a, nanocomposite films exhibited more improvement in scratch hardness with increasing filler content compared with polyamide films containing a dispersion of microfiller with a particle diameter of 3 μm.

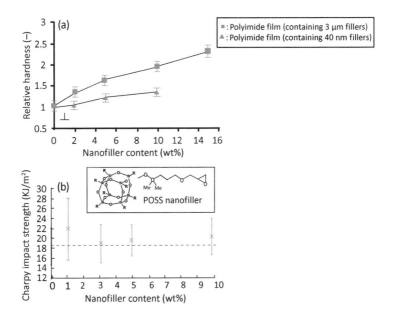

Figure 6.18 Mechanical properties of various kinds of nanocomposites [9, 10]. (a) Scratch hardness of polyamide nanocomposites. (b) Charpy impact strength of epoxy/POSS nanocomposites.

6.2.6.2 Charpy impact strength of epoxy resin/basket-shaped silica nanocomposites

A report showed the Charpy impact strength of epoxy resin nanocomposite dispersed with basket-shaped silica (polyhedral oligomeric silsesquioxane; POSS) nanofiller [10]. Charpy impact strength is measured by making a wedge-shaped cut (notch), fixing one side of the material, and striking the unfixed side with a hammer from the direction of the notch. As shown in Fig. 6.18b, the Charpy impact strength of epoxy resin nanocomposite dispersed with POSS nanofiller was about the same as that of the base epoxy resin, indicating little improvement in Charpy impact strength by the POSS nanofiller dispersion.

6.2.6.3 Mechanical characteristics of nano- and microcomposites

There have been reports on the mechanical characteristics of nano- and microcomposites, made with a high content of conventional microfiller (60–70 wt%) and a few weight percent

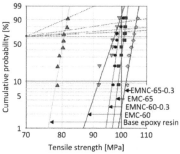

Base epoxy resin (without fillers)
EMC-60: Epoxy-based microcomposite (Micro-fillers: 60 wt%)
EMC-65: Epoxy-based microcomposite (Micro-fillers: 65 wt%)
EMNC-60-0.3: Epoxy-based nano- and microcomposite (Micro-fillers: 60 wt%, Nano-fillers: 0.3 wt%)
EMNC-65-0.3: Epoxy-based nano- and microcomposite (Micro-fillers: 65 wt%, Nano-fillers: 0.3 wt%)

(a) Weibull plots of tensile strength

Sample	Scale parameter (at 63.2 % probability) [MPa]	Shape parameter [−]
Base epoxy resin	80.6	78.4
EMC-60	97.3	39.6
EMC-65	101.7	78.7
EMNC-60-0.3	98.9	94.7
EMNC-65-0.3	105.9	104.6

Sample	Scale parameter (at 63.2 % probability) [MPa]	Shape parameter [−]
Base epoxy resin	107.6	65.1
EMC-60	157.8	24.4
EMC-65	171.2	31.7
EMNC-60-0.3	162.7	55.9
EMNC-65-0.3	176.9	65.0

(b) Weibull parameters of tensile strength   (c) Weibull parameters of bending strength

Figure 6.19 Mechanical properties of nanomicrocomposites. © 2011 IEEE. Reprinted, with permission, from Park, J., Lee, C., Lee, J., and Kim, H. (2011). Preparation of Epoxy/Micro- and Nano-composites by Electric Field Dispersion Process and Its Mechanical and Electrical Properties, *IEEE Trans. Dielectr. Electr. Insul.*, **18**(3), pp. 667–674.

of nanofiller [11]. Figure 6.19 is a Weibull distribution of the tensile strength of a nano- and microcomposite dispersed with silica microfiller (60 or 65 wt%) and a layered filler (0.3 wt% clay). Moreover, Fig. 6.19b shows the result of the tensile strength after statistical treatments. Nano- and microcomposites show a higher tensile strength and less scattering than microcomposites. Additionally, Fig. 6.19c shows an improved bending strength of nano- and microcomposites.

We have introduced several mechanical characteristics of nanocomposite insulating materials. For application to actual electric power apparatuses and electronic devices, insulators need to have sufficient insulating materials. and satisfactory mechanical characteristics. The benefit of nanocomposites is improved mechanical characteristics of polymers with only small amounts of dispersed nanofiller.

References

1. Toray Research Center (2002). *Technological Trend of Nano-Controlled Composite Materials*, Chapter 1.
2. Yasuda, T. (2000). *Plastics*, **51**, pp. 104–111.
3. Lan, T., and Pinnavaia, T. (1994). Clay-Reinforced Epoxy Nano-composites, *Am. Chem. Soc. Chem. Mater.*, **6**(12), pp. 2216–2219.
4. Huang, X., Xie, L., Jiang, P., and Liu, F. (2010). Enhancing the Permittivity, Thermal Conductivity and Mechanical Strength of Elastomer Composites by Using Surface Modified $BaTiO_3$ Nanoparticles, *Proc. IEEE ICSD*, No. G3-2, pp. 830–833.
5. Nose, J., Yamano, S., Kozako, M., Ohki, Y., Koutou, M., Okabe, N., and Tanaka, T. (2005). Preliminary Examination of Bending Characteristics of Epoxy/Alumina Nanocomposite Materials, *IEEJ National Convention Record*, No. 2–108, p. 123.
6. Imai, T., Sawa, F., Ozaki, T., Nakano, T., Shimizu, T., and Yoshimitsu, T. (2004). Preparation and Insulation Properties of Epoxy-Layered Silicate Nanocomposite, *IEEJ Trans. FM*, **124**(11), pp. 1065–1072.
7. Imai, T., Sawa, F., Ozaki, T., Shimizu, T., Kuge, S., Kozako M., and Tanaka, T. (2006). Effects of Epoxy/Filler Interface on Properties of Nano- or Micro-composites, *IEEJ Trans. FM*, **126**(2), pp. 84–91.
8. Ochi, K., Harada, M., Minamikawa, S., and Nakayama, K. (2005). Thermo-Mechanical Properties of Nano-Composites Prepared Form Silsesqioxane-Type Epoxy Resins and Layered Titanate, *IEEJ The*

Paper of Technical Meeting on Dielectrics and Electrical Insulation, No. DEI-05-84, pp. 23–28.

9. Irwin, P. C., Cao, Y., Bansal, A., and Schadler, L. A. (2003). Thermal and Mechanical Properties of Polyimide Nanocomposites, *Annual Rept. IEEE CEIDP*, pp. 120–123.

10. Takala, M., Karttunen, M., Pelto, J., Salovaara, P., Munter, T., Honkanen, M., Auletta, T., and Kannus, K. (2008). Thermal, Mechanical and Dielectric Properties of Nanostructured Epoxy-Polyhedral Oligomeric Silsesquioxane Composites, *IEEE Trans. Dielectr. Electr. Insul.*, **15**(5), pp. 1224–1235.

11. Park, J., Lee, C., Lee, J., and Kim, H. (2011). Preparation of Epoxy/ Micro- and Nano-composites by Electric Field Dispersion Process and Its Mechanical and Electrical Properties, *IEEE Trans. Dielectr. Electr. Insul.*, **18**(3), pp. 667–674.

6.3 Long-Term Characteristics

Toshio Shimizu

Power and Industrial Systems R&D Center, Toshiba Corporation, Fuchu-shi, Tokyo 183-8511, Japan

toshio4.shimizu@toshiba.co.jp

When developed materials are applied to products, it is important to ensure that they also have excellent long-term characteristics. In electric power apparatuses and electric devices, it is necessary to verify long-term thermal stability and mechanical properties in addition to the electrical insulation properties discussed in the foregoing chapter. Polymer nanocomposite technology is also effective in improving long-term characteristics.

6.3.1 Thermal Endurance of Polymers Is Improved by Nanocomposites

Organic polymers deteriorate more under UV, radiation and thermal environments than metals and ceramics. Electric power apparatuses and electric devices are exposed to temperature rise in operation; so it is important to evaluate their performance under these conditions. Actually, most material properties of polymers, such as mechanical, electrical, and physical properties, change dramatically over the glass transition temperature, so the temperature of materials must be regulated. As well as short-term evaluation for temperature, it is necessary to evaluate long-term thermal properties to satisfy the design life of actual products.

Successive deterioration under hot conditions is mainly caused by oxidation and decomposition of polymers. Because these are chemical reactions, their velocity depends on temperature and oxygen density and progresses with time. Generally, in the case of polymers, molecular chains break into small molecules, thus decreasing weight, mechanical strength, and elongation, causing deterioration of material characteristics. The thermal endurance of a polymer is evaluated by an acceleration examination under high temperature conditions. Parameters evaluated are change of weight, mechanical strength, and elongation with time.

For example, Fig. 6.20 shows a weight change when isotactic polypropylene (iPP) was held in an oven at a constant temperature of 110°C [1]. Polypropylene suddenly decomposes after a certain time has passed, and its weight decreases. This decomposition can be delayed by adding clay 6% by weight as a nanofiller. Furthermore, because it is thought that this characteristic drop depends on the chemical reaction velocity, it is possible to arrange it in a plot of the Arrhenius form and evaluate it in a short time. Oxidation and decomposition reactions can be restrained by a nanocomposite filled with clay, thus improving the thermal endurance.

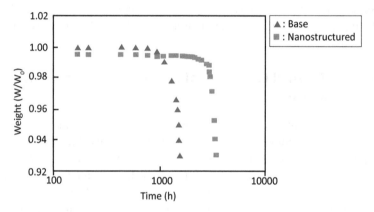

Figure 6.20 Weight vs. time plots obtained from conventional accelerated thermal endurance test, carried out at 110°C on base and nanostructured iPP [1].

As explained above, one of the factors in the thermal deterioration of the polymer is the existence of oxygen. It is possible to reduce deterioration by controlling contact between polymer and oxygen. Clay exfoliated by an appropriate method has a large aspect ratio, so the surface area is very large. As shown in Fig. 6.21, even with a small addition, the clay disperses closely in the polymer and decreases the exposure of the polymer to oxygen. It also controls the oxidative degradation in the polymer by suppressing the diffusion of oxygen molecules to the polymer inside, and developing so-called gas barrier characteristics, and it is thought that this can improve thermal endurance. A nanocomposite shows improved gas barrier characteristics in many plastic films, and it has been applied to plastic bottles of carbonated water.

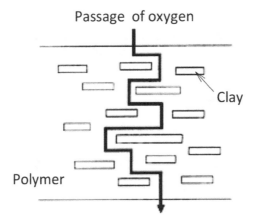

Figure 6.21 Gas barrier effect caused by adding nanofiller.

6.3.2 Fatigue Properties of Polymer Are Improved by Nanocomposite

In designing a product, creep and fatigue properties are important long-term mechanical characteristics. Like metallic materials, if mechanical external force and transformation (strain) are added to a polymer for a long time, it will gradually transform and finally lead to destruction. Creep is characterized as deformation change with time or time to rupture under constant external force or constant transformation. On the other hand, fatigue is defined as deformation change with repetition number or number of cycles to rupture when external force or transformation fluctuates. In this clause, the effect of nanocomposite on fatigue properties is introduced.

Fatigue is strength under fluctuating load such as expansion and shrinkage by vibration, temperature change, or pressure change. It is evaluated by number of repetitions before fracture under sine wave repetition stress or distortion of a specimen at a specified frequency using testing equipment. Because the mechanical fracture modes of materials are different under static load and repetition load, it is important to confirm the effects of nanocomposite on both characteristics.

For example, Fig. 6.22 shows the temperature effects on tensile strength of materials with added clay (2% (NCH-2) or 5% (NCH-5)) by weight for nylon 6 [2]. The tensile strength of nylon 6 decreases with temperature rise. The static mechanical strength was improved by adding a small amount of clay. However, the influence of the quantity of added clay is not confirmed.

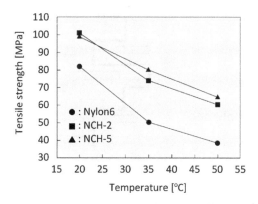

Figure 6.22 Temperature effects on tensile strength of nylon 6 [2].

Figure 6.23 shows fatigue properties of the same materials at normal temperature. Several examinations were carried out by changing the stress level (maximum tensile stress). Maximum stress was plotted on a vertical axis and number of cycles to failure was plotted as logarithm indication on a cross axle. If the stress level is high, fatigue life is short, and fatigue life increases as stress decreases. This is a representative diagram indicating fatigue properties, called the S-N (Stress-Number) curve. Like tensile strength, fatigue properties were improved by adding clay to nylon 6 at normal temperature. The fatigue properties of the same materials at 35°C are shown in Fig. 6.24. Nylon 6 not filled with clay showed greatly decreased fatigue properties. The strength of the polymer itself decreased with temperature rise, and this changed the fracture mode. On the other hand, the deterioration of fatigue properties of materials filled with clay was small. Clay in a polymer functioned as a reinforcing structure, and effectively suppressed the strength drop of the polymer with temperature rise. In addition, a difference in fatigue properties was observed with increased clay filling, and 5% by weight addition was most effective.

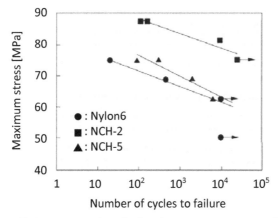

Figure 6.23 Fatigue properties of nylon 6 at room temperature [2].

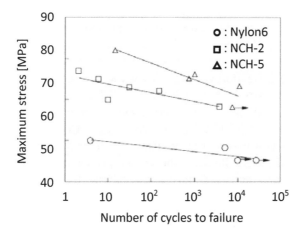

Figure 6.24 Fatigue properties of nylon 6 at 35°C [2].

The effect of nanocomposite on fatigue properties was also confirmed in a typical thermosetting epoxy resin [3]. Because epoxy resin cross-links in three dimensions, it is brittle under glass transition temperature, and is classified as a hard and brittle material. When mechanical load increases, unlike other thermoplastic resins, epoxy resin ruptures without large transformation. Therefore, various kinds of particles have been tried as fillers to increase transformation until rupture and to improve mechanical strength. For example, fatigue properties have been reported for epoxy resin filled with rubber (CTBN:

carboxyl-terminated butadiene-nitrile rubber) and with nanosilica together with rubber [4]. In this report, it is shown for neat epoxy resin that fatigue properties were improved by adding rubber and nanosilica. The nanosilica debonded from the resin during a fatigue test on the epoxy resin, and more energy was required to cause plastic deformation, thus improving the resin's fatigue properties. Rubber also causes plastic deformation by cavitation. In addition, the epoxy resin filled with both rubber and nanosilica showed superior fatigue characteristic by synergy. Because epoxy resin is applied widely, for example, as matrix resin for FRP (fiber-reinforced plastics), it requires characteristic as a structural element; so improved fatigue characteristic are also very important.

One of the progress modes of damage due to fatigue is crack propagation. Like metallic materials, a crack in a polymer propagates little by little with repeated transformation by fatigue, and if crack length reaches a critical value, rupture occurs. Figure 6.25 shows fatigue characteristics as a relationship between propagation velocity of crack length and stress level [5]. The vertical axis is crack propagation velocity and is the change of crack length per load reversal measured with a microscope during repeated loading. The horizontal axis ΔK is stress intensity factor and is a mechanical intensive factor as a function of crack length and stress. Because the cracks propagated gradually and their lengths increased during the fatigue test under constant stress, both ΔK and crack propagation velocity increased. For fatigue crack propagation characteristics for materials with added carbon black (CB), carbon nanotube (CNT) and a vapor phase epitaxy carbon fiber (VGCF) to epoxy resin of 1% by weight as a nanofiller, the curves shifted to the right side, and crack propagation was delayed for the same stress level. Particularly, with the addition of fiber-formed nanofillers such as CNT and VGCF, the improvement of fatigue crack properties was confirmed. The scanning electron microscope (SEM) photograph of the fracture surface after the examination is shown in Fig. 6.26. The fracture surface of the neat epoxy resin was smooth, showing that crack propagation resistance was small. On the other hand, in the materials with added nanofillers, a hoof-formed pattern was observed. This was caused by plastic deformation behind the fillers in the crack propagation direction and was confirmed only around nanofillers. Therefore, like the said article, filling with nanofillers caused partial transformation to the epoxy resin, which was hard to transform and it needed a lot of energy for destruction. There was a step in this hoof part, but the step was larger in CNT and VGCF than in CB, and energy consumption increased because the crack was branched off by the form of the fiber nanofiller, and crack propagation was restrained. Furthermore, it was reported that the fatigue crack propagation characteristic was improved by the addition of nanosilica [6].

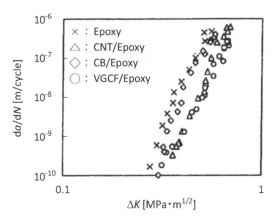

Figure 6.25 Fatigue crack propagation properties of epoxy resin [5].

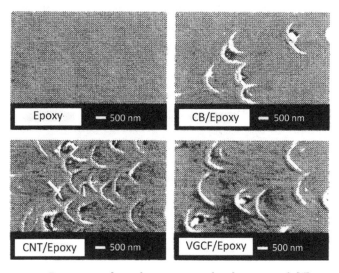

Figure 6.26 Fracture surface of epoxy resin after fatigue crack [5].

The fatigue fracture mode of a polymer is changed by filling a polymer with a nanofiller as described above and a lot of energy is required for crack propagation. It is very important for nanocomposites to improve the fatigue properties of the polymer as well as insulation properties for long-term reliability of apparatuses.

References

1. Saccani, A., Motori, A., Patuelli, F., and Montanario, G. C. (2007). Thermal Endurance Evaluation of Isotactic Poly(propylene) Based Nanocomposites by Short-term Analytical Methods, *IEEE Trans. Dielectr. Electr. Insul.*, **14**(3), pp. 689–695.

2. Kichise, M., Shijie, Z., Usuki, A., and Kato, M. (2011). Temperature Influence on Fatigue Fracture of Nylon 6 Clay Hybrid Composite Materials, *J. Soc. Mater. Sci. Japan.*, **60**(5), pp. 457–463 (in Japanese).

3. Juwono, A., and Edward, G. (2006). Mechanism of Fatigue Failure of Clay-Epoxy Nanocomposites, *J. Nanosci. Nanotechnol.*, **6**, pp. 3943–3946.

4. Manjunatha, C. M., Taylor, A. C., Kinloch, A. J., and Sprenger, A. (2009). The Cyclic-Fatigue Behaviour of an Epoxy Polymer Modified with Micron-Rubber and Nano-Silica Particles, *J. Mater. Sci.*, **44**, pp. 4487–4490.

5. Utsumi, S., Matsuda, S., and Kishi, H. (2008). Effect of Nanofiller on the Fatigue Property, *Sci. Counc. Japan Proc. 52nd Japan Cong. Mater. Res.*, **42**, pp. 287–288 (in Japanese).

6. Blackman, B. R. K., Kinloch, A. J., Sohn Lee, J., Taylor, A. C., Agarwal, R., Schueneman, G., and Sprenger, S. (2007). The Fracture and Fatigue Behaviour of Nano-Modified Epoxy Polymer, *J. Mater. Sci.*, **42**, pp. 7049–7051.

PART 4
THEORETICAL ASPECTS

Chapter 7

Structures of Polymer/Nanofiller Interfaces

Toshikatsu Tanaka[a] and Muneaki Kurimoto[b]

[a]*Waseda University*
[b]*Nagoya University*

7.1 Interfaces Have Volume

Toshikatsu Tanaka

IPS Research Center, Waseda University,
Kitakyushu-shi, Fukuoka 808-0135, Japan

t-tanaka@waseda.jp

Polymer nanocomposites are innovative in their inherent characteristics of "interfaces" between nanofillers and their surrounding polymer matrices. Various models are proposed for such interfaces. A bound polymer model is an excellent concept derived from colloid chemistry. This indicates that nanofillers are surrounded and tightly bound with polymer chains. Change in glass transition temperatures leads to a two-layer model. A multi-core model is devised based on quantum mechanics. All the models stated above support the contention that interfaces of polymers with nanofillers have thickness and volume.

Advanced Nanodielectrics: Fundamentals and Applications
Edited by Toshikatsu Tanaka and Takahiro Imai
Copyright © 2017 Pan Stanford Publishing Pte. Ltd.
ISBN 978-981-4745-02-4 (Hardcover), 978-1-315-23074-0 (eBook)
www.panstanford.com

7.1.1 What Are Interfaces?

Polymer nanocomposites are defined as composite materials constituted of inorganic nanofillers and organic polymers, where at least one dimension of nanofillers must be in the range of 1–100 nm. Specific surfaces of nanofillers are enormous compared to those of microfillers, which have been used for a long time. In that sense, interfaces formed between the two kinds of substances exert a big influence on the overall characteristics of nanocomposites [1]. Chemical bonding reaction is not likely to take place between organic and inorganic species, but it can be done with the help of binding agents like coupling agents. Structures of interfaces are complex and attractive. Several kinds of physical and chemical models are proposed to help understand various characteristic phenomena that nanocomposites exhibit. Some of the schematic interfacial models are demonstrated in Fig. 1.1, Chapter 1. The interfaces are different in structure from either base polymers or added nanofillers. Interfaces with thickness are either of random structure or of spherulite structure, as shown in Figs. 1.1a,b. Furthermore, a model with formed multiple different layers is demonstrated in Fig. 1.1c.

7.1.2 Interfaces Formed by Inorganic Fillers and Organic Polymers: What Features Do They Have?

7.1.2.1 Interfaces account for enormous rate in nanocomposites

Interfaces are crucial in polymer nanocomposites. Their performance is determined by many different characteristic parameters such as the shape, size, and loading ratio of the nanofillers, inter-filler distance, interfacial morphology, and even mesoscopic structures. Nanofillers are spherical, oblate-spheroidal, platelet, layered, or acicular in shape. Silica, titania, alumina, magnesia, and layered silicate investigated in the electrical insulation field are of characteristic shape based on crystallographic properties. An inter-filler distance is, for instance, 70 nm (surface to surface), when nanofillers of 40 nm in size are filled only by 5 wt%. A total surface becomes even 3.5 km^2/m^2 under the same condition. On the other hand, when microfillers 100 µm in diameter are loaded by 10 wt%, the inter-filler distance and the sum of interfacial surfaces are 122 µm and 0.00289 km^2/m^2. The ratio is 1.21×10^3. A big difference is recognized in the inter-filler distance (three times) and the summed surfaces (three order times). These results clearly state how important the interfaces are [2]. Therefore, it can be said that mesoscopic characteristics originating from organic-inorganic interfaces emerge in macroscopic performance of polymer nanocomposites.

7.1.2.2 Silane couplings fuse inorganic and organic matters into polymer nanocomposites

Silane couplings are typical bonding states that combine organic polymers with inorganic fillers. Bonding strength is medium (bonding energy: 5 to 10 kcal/mol) in hydrogen bonds, and becomes stronger when the chemical bonding is converted from a hydrogen bond to a covalent bond. Organic polymers and inorganic fillers are immiscible with each other, so coupling agents like silane combine them with the help of hydrogen bonding, as shown in Fig. 7.1. Strong ionic bonding and weak van der Waals bonding are also effective in integrating them. "Wettability" is also an important factor and plays a functional role in enhancing affinity by which two kinds of substances are adjacent to each other.

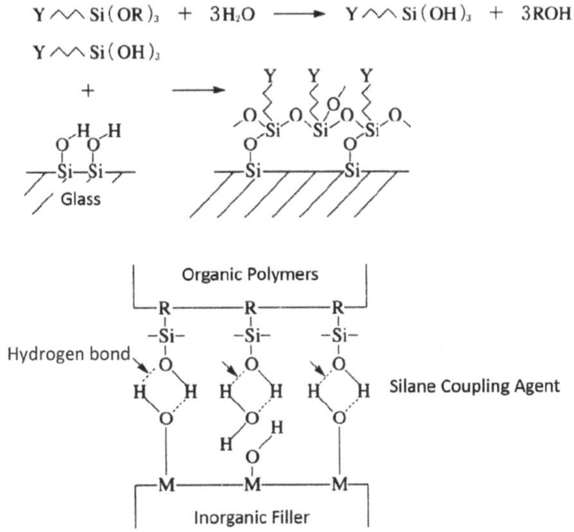

Figure 7.1 Chemical structures of silane coupling.

7.1.2.3 What are interfacial interactions like?

Interaction forces originate from chemical bonding (covalent bonds and ionic bonds), hydrogen bonding, van der Waals force, magnetic forces, and mechanical forces as anchor effects. Mechanical interactions are represented by a concept of "bound polymers" [3], as schematically shown in Fig. 7.2. This figure depicts that an immobile or bound layer is formed around a nanofiller, since matrix polymer molecules or chains are constrained from free movement in the neighborhood of the nanofiller. As interfacial thickness is considered to increase with increase of interaction strength, it is a possible measure for interaction strength, since it can be estimated from experiments. As a result, it is shown to be in the range of 10 to 200 nm in the composite system of polyvinyl chloride (PVC) with inorganic fillers. When interaction zones are formed, physical changes follow, as indicated below [4]:

(i) Mobility of polymer chains changes.
(ii) Free volume changes.
(iii) Glass transition temperature changes.
(iv) Inner and inter spherulite structures change.
(v) Percolation takes place if adjacent nanofillers contact each other.

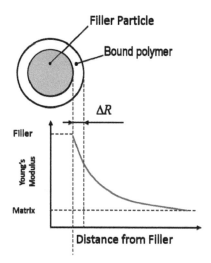

Figure 7.2 Schematics of a bound polymer layer on filler surface [3].

7.1.3 Various Interfacial Models Are Proposed

7.1.3.1 A simple two-layered interfacial model [4]

Glass transition temperature (T_g) is lowered or heightened depending on the strength of polymer-nanofiller interaction. Some experiments were carried out on a thin polymer film fabricated on a flat inorganic plate, which simulates an interface around a nanofiller on a flat surface converted from its original curved surface. The following results are obtained from this thin film study:

(i) Glass transition temperature declines when the inner side of an interface is non-wet and/or the outer side forms a free surface without any interaction.
(ii) Glass transition temperatures do not decline when both sides of the thin film are sandwiched by two separate substrates.
(iii) Glass transition temperatures possibly increase when interfaces are wet as interfacial interaction is strong.

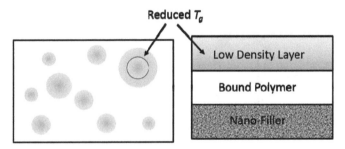

Figure 7.3 Two-layer interface model [4].

A two-layered interfacial model as shown in Fig. 7.3 is depicted to support and explain the above statement. Thickness is supposed to be in the range of several nm to several hundred nm. The model consists of two layers with different characteristics, i.e., a proximity layer in contact with bulk polymers and an adjacent layer touching a substrate (the surface of a nanofiller). It was well accepted that glass transition temperatures are lower in the proximity layer than in the polymer matrix, and constant over its inside. However, recently it was found that glass transition temperatures show a gradient from the proximity side to the polymer matrix side. Furthermore, a revised model is proposed in which the proximity layer constitutes a mobile liquefied layer. In such a case, the conductivity of ions and the transmittance of organic molecules are expected to increase. Interactions at interfaces and between nanofillers and polymer matrices remain to be solved.

7.1.3.2 Multi-core model analogously derived from quantum mechanics and collide chemistry [5]

A concept of "interaction zones" is proposed based on the fact that interfacial characteristics emerge in the macroscopic performance of polymer nanocomposites, since the fraction that interfaces occupy in the materials is enormous. A multi-core model shown in Fig. 7.4 is derived to clarify their fine structures. According to this model, spherical nanofillers with dimension of several tens of nanometers in diameter are homogeneously dispersed and separated in similar geometrical scale. Generally inter-filler distance is 1 to 10 times the filler diameter. If this distance approaches 1, inter-filler interactions possibly emerge.

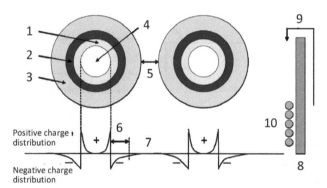

1: The first layer with mutual tight binding: several nm
2: The second layer with deep trap: about 10 nm
3: The third layer with large local free volume as ion traps and shallow electronic gaps: several tens nm
4: Nano-particles: 20 - 40 nm in diameter
5: Inter-filler distance (surface to surface): 40 - 100 nm
6: Debye shielding distance: up to about 100 nm
7: Possible overlapping of the third layers and charge tails of nano-fillers
8: Electrode facing accumulated charge tails of nano-fillers
9: Charge carrier injection via Schottky emission at high electric field
10: Collective charge tail effect will modify carrier injection

Figure 7.4 Multi-core interface model [5].

It must be borne in mind that interfacial zones have different characteristics than either polymer matrices or nanofillers. Considering the various effects of interfaces on dielectric properties clarified so far, a multi-core model with only a threefold structure (actually a three-core model) is analogously derived based on the knowledge of quantum mechanics and collide chemistry. Each of the three cores is briefly explained below:

(i) First layer (bonded layer)

This inside layer belongs to a region in which inorganics are chemically bonded to organics by ionic bond, covalent bond, hydrogen bond, or van der Waals force. Both organics and inorganics are tightly bonded with each other via silane coupling agents in excellent polymer nanocomposites. Observation of small-angle X-ray diffraction (SAXS) data indicates that silica bleeds partially into polymer matrices in the case of covalent bond, and is mutually mixed with polymer chains to form a network in the case of hydrogen bond.

(ii) Second layer (bound layer)

This middle layer is represented by a region consisting of morphologically ordered structures like spherurites formed under the influence of the first layer formation process. The forming process of this second layer is closely related to polymer chain mobility (glass transition temperatures and stereographic conformation) and crystallinity (selective adsorption to nanofillers during curing).

(iii) Third layer (loose layer)

This outer layer is formed under the influence of the second layer formation process. For example, in the case of epoxy resins, the third layer is created around the second layer as "a curing-agent-depleted region," i.e., "a stoichiometrically unsatisfied region," to form a layer with lower density.

The charge is transferred due to the difference between electrochemical potentials. For this reason, an electric double layer like the Gouy–Chapman diffuse layer is formed, giving rise to a peculiar influence on dielectric and insulation characteristics. A rule of thumb on triboelectricity leads to the fact that polyethylene (PE), polypropylene (PP), ethylene vinyl acetate

(EVA) copolymers and the like are negatively charge against metal, while silicone elastomers, polyamide, epoxy and the like are positively charged. Coulomb's force generated under such conditions is a far-distant force. Since the inter-filler distance is only 2 to 3 times the nanofiller diameter, next filler neighbors electrically interact with each other. This may induce electrostatically interacting cooperative phenomena such as percolation effect.

Nanofillers in polymers are macroscopically neutral but electrostatically charged inside Debye length. The electric field induced by this effect is considered to influence the transit of electrons. Electrons gain high energy in the treeing initiation stage. Resulting high-energy electrons may induce Coulombic interactions with the field created by nanofillers to be closely involved in tree initiation and growth phenomena. Various explanations of this model and new model proposals as well as computer simulations are under way [6–13].

7.1.3.3 Water shell model devised for interfaces with minute gaps [14]

Water is detrimental in polymer nanocomposites when it is contained in defects like voids between nanofillers and polymer matrices. This is based on a series of experiments on dielectric properties carried out on the effects of absorbed water. Specimens consisting of base epoxy, microsilica (40 μm in size) filled epoxy, and nanosilica (50 nm in size) added epoxy are prepared. Relative humidity is adjusted at room temperature to 353 K by using a saturated solution of salt in water. The frequency for dielectric spectroscopy ranged from 10^{-3} to 10^5 Hz. A hydration isothermal line (i.e., water intake quantity) is determined by measuring the mass as a function of relative humidity. As a result, it was elucidated that the nanocomposite absorbs 60% more water than the base epoxy and the microcomposite. From dielectric spectroscopy, some difference is recognized in electrical conductivity and quasi-Davidson–Cole behaviors of the nanocomposite in the frequency region below 10^{-2} Hz. That is to say, activation energy is influenced by hydration and temperature. From the above observation, a water shell model is derived as shown in Fig. 7.5 [14].

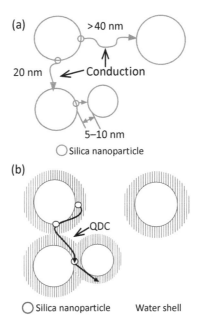

Figure 7.5 Water shell model of interfaces [14].

Under dry conditions, the neighboring shells are never overlapped, as shown in Fig. 7.5a. Under wet conditions, the neighboring shells are possibly overlapped by forming water shells, as shown in Fig. 7.5b. If there is excess water, it degrades interfaces and connects neighboring nanofillers together one after another. Probable percolation and/or partial breakdown may lead to total breakdown of the specimen. Therefore, interfaces should be chemically modified to have hydrophobic characteristics.

References

1. Lewis, T. J. (2004). Interfaces Are the Dominant Feature of Dielectrics at the Nanometric Level. *IEEE Trans. Dielectr. Electr. Insul.*, **11**(5), pp. 739–753.

2. Ajayan, P. M., Schadler, L. S., and Braun, P. V. (2003). *Nanocomposite Science and Technology* (WILEY-VCH Verlag GmbH & Co. KGaA, Weinheim, Germany).

3. Filler Research Society, ed. (1994). *Composite Materials and Fillers*, (CMC Press, in Japan).

4. Mays, A. M. (2005). Nanocomposites: Softer at the Boundary, *Nat. Mater.*, **4**, Sept., pp. 651–652.

5. Tanaka, T., Kozako, M., Fuse, N., and Ohki, Y. (2005). Proposal of a Multi-core Model for Polymer Nanocomposite Dielectrics. *IEEE Trans. Dieletr. Electr. Insul.*, **12**(4), pp. 669–681.

6. Tanaka, T. (2006). Interpretation of Several Key Phenomena Peculiar to Nano Dielectrics in terms of a Multi-core Model, *Annual Rept. IEEE CEIDP*, No. 3B-2, pp. 298–301.

7. Su, Z., Li, X., and Yin, Y. (2011). The Three-Layered Core Model Permittivity of Polymer Nano-composites in Electrostatic Field, *Proc. IEEJ ISEIM*, No. MVP 1–19, pp. 297–300.

8. Shi, N., and Ramprasad, R. (2008). Local Properties at Interfaces in Nanodielectrics: An ab initio Computational Study, *IEEE Trans. Dielectr. Electr. Insul.*, **17**(1), pp. 170–177.

9. Andritsch, T., Kochetov, R., Morshuis, P. H. F., and Smit, J. J. (2011). Proposal of the Polymer Chain Alignment Model, *Annual Rept. IEEE CEIDP*, **2**(6-1), pp. 624–627.

10. Kuehn, M., and Liem, H. K. (2004). Simulating Nanodielectric Composites using the Method of Local Fields, *Annual Rept. IEEE CEIDP*, No. 4-1, pp. 310–313.

11. Sawa, F., Imai, T., Ozaki, T., Shimizu, T., and Tanaka, T. (2007). Molecular Dynamics Simulation of Characteristics of Polymer Matrices in Nanocomposites, *Annual Rept. IEEE CEIDP*, No. P3-11, pp. 263–266.

12. Sawa, F., Imai, T., Ozaki, T., Shimizu, T., and Tanaka, T. (2009). Coarse Grained Molecular Dynamics Simulation of Thermosetting Resins in Nanocomposite, *Proc. IEEE ICPADM*, No. H-32, pp. 853–856.

13. Smith, J. S., Bedrov, D., and Smith, G. D. (2003). A Molecular Dynamics Simulation Study of Nanoparticle Interactions in a Model Polymer-nanoparticle Composite, *Compos. Sci. Technol.*, No. 63, pp. 1599–1605.

14. Zou, C., Fothergill, J. C., and S W. Rowe, S. W. (2008). The Effect of Water Absorption on the Dielectric Properties of Epoxy Nano-composites, *IEEE Trans. Dielectr. Electr. Insul.*, **15**(1), pp. 106–117.

7.2 Physicochemical Analysis Methods for Interfaces

Muneaki Kurimoto

Department of Electrical Engineering and Computer Science,
Nagoya University, Nagoya-shi, Aichi 464-8603, Japan

kurimoto@nuee.nagoya-u.ac.jp

There are a large number of interfaces in a polymer nanocomposite. The interfaces are characterized by the shape and dimensions of the nanofiller, the amount of the added nanofiller (filling amount), the distance between nanofillers, and the bonding states of organic and inorganic substances. In addition, nanofiller agglomerates inhibit the formation of interfaces. How can the properties that characterize three-dimensional interfaces be evaluated? Physicochemical methods used for the evaluation are explained below.

7.2.1 Shapes, Sizes and Dispersion Are Evaluated by SEM and TEM Observation

The shape and dimensions of the nanofiller are evaluated by light scattering or laser diffraction in a liquid and electron microscopy. In particular, the nanofiller in cured polymers is generally evaluated by electron microscopy. In electron microscopy, the difference in material composition is represented as image contrast on the basis of the information on electrons emitted from the material irradiated with electron beams. Therefore, the shape and dimensions of the nanofiller with a size of several nanometers to several millimeters can be evaluated [1].

Scanning electron microscopy (SEM) is a type of electron microscopy for visualizing the material composition in the vicinity of the material surface using secondary electrons emitted from the surface. There are various methods of preparing the surface to be observed, for example, obtaining a smooth fracture surface by breaking a specimen frozen in liquid nitrogen and surface polishing. In addition, conductivity is generally applied to the surface of an insulating material by coating the surface with carbon or osmium so that electrons in electron beams do not accumulate on the surface. Figure 7.6 shows the SEM images of a nanocomposite obtained by adding an alumina nanofiller to epoxy resin. The surface is polished using an ion gun, coated with osmium, and then observed by field emission SEM (FE-SEM) [2]. Spherical alumina nanofiller particles with a size of several tens of nanometers and nanofiller agglomerates with a size of several tens of micrometers are clearly observed.

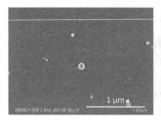 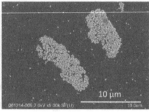

Figure 7.6 SEM images of a nanocomposite obtained by adding an alumina nanofiller to epoxy resin.

Transmission electron microscopy (TEM) is a type of electron microscopy for visualizing the composition of materials in a specimen using electrons transmitted through the specimen. Electrons transmitted through a thin specimen sliced to a thickness of 20–100 nm are observed. Figure 7.7 shows the TEM image of a nanocomposite obtained by adding layered silicate (clay) to epoxy resin [3]. As shown in the figure, layered silicate with a thickness of 2–3 nm is confirmed. Generally, TEM has a higher resolution than SEM and enables the observation of the inside of a specimen.

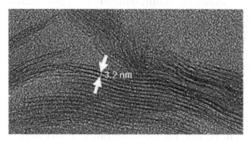

Figure 7.7 TEM image of a nanocomposite obtained by adding layered silicate (clay) to epoxy resin. © 2011 IEEE. Reprinted, with permission, from Park, J., Lee, C., Lee, J., and Kim, H. (2011). Preparation of Epoxy/Micro- and Nano-Composites by Electric Field Dispersion Process and Its Mechanical and Electrical Properties, *IEEE Trans. Dielectr. Electr. Insul.*, **18**(3), pp. 667–674.

When a specimen is irradiated with electron beams during electron microscopy, the characteristic X-rays originating from the elements in the specimen are emitted. Energy-dispersive X-ray spectroscopy (EDX or EDS) enables the elemental and compositional analyses of the area to be observed by detecting characteristic X-rays, which is a function of electron microscopy. Figure 7.8 shows the (a) TEM image and [(b) and (c)] EDS spectra of a nanomicrocomposite obtained by adding layered silicate and microsilica fillers to epoxy resin [4]. Figure 7.8b shows the EDS spectrum of the characteristic X-rays emitted from a microparticle shown in Fig. 7.8a. The characteristic X-rays originating from Si and O are observed, indicating that the microparticle is a SiO_2 particle. Figure 7.8c shows the EDS spectrum of the characteristic X-rays emitted from the lower region in Fig. 7.8a. The characteristic X-rays originating from Si, O, Al, Fe, and Mg are seen, and layered silicate is observed in this region.

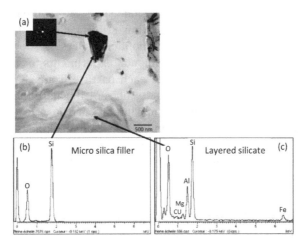

Figure 7.8 TEM image (a) and EDS spectra (b), (c) of a nanomicrocomposite obtained by adding layered silicate and microsilica fillers to epoxy resin. © 2008 IEEE. Reprinted, with permission, from Fréchette, M. F., Larocque, R. Y., Trudeau, M., Veillette, R., Rioux, R., Pélissou, S., Besner, S., Javan, M., Cole, K., Ton That, M.-T., Desgagnés, D., Castellon, J., Agnel, S., Toureille, A., and Platbrood, G. (2008). Nano-Structured Polymer Microcomposites: A Distinct Class of Insulating Materials, *IEEE Trans. Dielectr. Electr. Insul.*, **15**(1), pp. 90–105.

Researchers have attempted to process the SEM and TEM images of nanocomposite insulation materials to analyze the grain size distribution of nanofiller particles or agglomerates [2, 5, 6]. The following points should be considered in this analysis.

(i) The number of particles to be observed should be increased by increasing the number of target areas in order to analyze the grain size distribution of the entire specimen, because only the local field is observed by one measurement (problem of sampling).

(ii) SEM provides two-dimensional information on particles in the vicinity of the surface of a specimen, whereas TEM provides two-dimensional information without depth information on particles, although the particles inside a specimen can be observed (problem of dimensions to be observed).

For the above reasons, the results of processing the SEM and TEM images, unlike the three-dimensional grain size distribution, can be used to evaluate the relative differences in grain size distribution and particle dispersion state by comparing the materials using the same evaluation criteria. Such information is useful when selecting a nanofiller-dispersing device and appropriate processes [5]. Fundamental experiments for obtaining three-dimensional information of each particle from SEM images have been carried out. The results of centrifugal sedimentation experiments clarified that the maximum particle size observed in SEM images is close to that in the three-dimensional grain size distribution [7]. This method can be used as a simple evaluation method for agglomerates that serve as defects in electrical insulation.

7.2.2 Filler Content Is Estimated by Measuring the Density of Nanocomposites

The filling amount of the nanofiller in nanocomposites is expressed by various units. Typical units include the volume ratio V_f (vol%), the weight ratio D_f (wt%), and parts per hundred of resin P [phr, the amount (g) of the nanofiller contained in 100 g of the polymer]. In the calculation of P, the weight of the curing agent is included in some cases but not in other cases. Note that P may differ twofold between these two cases. The filling amount of nanofiller in nanocomposites is mainly calculated from the amounts of the polymer and nanofiller measured during the manufacturing of the nanocomposites. However, the actual and measured filling amounts in a cured nanocomposite differ when the process of separating agglomerates is included in the manufacturing processes or when nanofiller particles are precipitated or scattered.

The filling amounts of nanofiller, V_f and D_f, can be precisely determined using the measured density of the nanocomposite. ρ_c is the density of the nanocomposite, ρ_e (g/mL) is the specific gravity of the base polymer, and ρ_f (g/mL) is the specific gravity of the nanofiller. These values are calculated as the ratio of the measured weight of each material in air to the volume calculated from the weight of each material in liquid (e.g., water and ethanol).

Assuming that the sum of the weights of the nanofiller and base polymer is 1 g in a specimen,

$$\frac{1 \times (1/\rho_c) \times (V_f/100)}{1/\rho_f} + \frac{1 \times (1/\rho_c) \times (1 - V_f/100)}{1/\rho_e} = 1\,g \qquad (7.1)$$

V_f is calculated using

$$V_f = \frac{\rho_c - \rho_e}{\rho_f - \rho_e} \times 100 \ [\text{vol\%}] \qquad (7.2)$$

Assuming that the sum of the volumes of the nanofiller and base polymer is 1 mL in a specimen,

$$\frac{1 \times \rho_c \times (D_f/100)}{\rho_f} + \frac{1 \times \rho_c \times (1 - D_f/100)}{\rho_e} = 1\,mL \qquad (7.3)$$

D_f is calculated using

$$D_f = \frac{1/\rho_e - 1/\rho_c}{1/\rho_e - 1/\rho_f} \times 100 = V_f \times \frac{\rho_f}{\rho_c} \ [\text{wt\%}] \qquad (7.4)$$

7.2.3 Inter-Filler Distances Are Evaluated in Mesoscopic and Macroscopic Scales

A quantitative evaluation technique that combines TEM and electron tomography (ET) is used to quantitatively evaluate the three-dimensional grain size distribution of nanofiller particles in a cured nanocomposite [8]. The principle of this technique is the same as that of computed tomography (CT). Several TEM images obtained from different observation angles are combined to calculate the three-dimensional position of each nanofiller to obtain the three-dimensional distribution of nanofiller particles.

Figure 7.9a shows the three-dimensional distribution of nanofiller particles in a nanocomposite obtained by adding a 17.4 wt% silica nanofiller to epoxy resin using TEM/ET. The three-dimensional positions of 1284 nanofiller particles in a target volume of 594 × 596 × 104 nm^3 are shown. The distance between nanofiller particles is determined mesoscopically (microscopically from nanometer to submicrometer order). Figure 7.9b shows the frequency distribution of the minimum distance between nanofiller particles in the nanocomposite obtained by adding a 10–17.4 wt% silica nanofiller to epoxy resin. With increasing amount of nanofiller, the distribution of the minimum distance between nanofiller particles shifts to smaller distances, indicating that the distance between nanofiller particles decreases with their increasing number density.

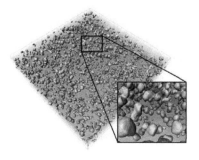
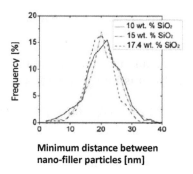

(a) Three-dimensional distribution of nano-fillers (17.4 wt%)

(b) Distribution of minimum distance between nano-fillers

Figure 7.9 Three-dimensional distribution and particle distance of nanofillers in a nanocomposite obtained by adding a 17.4 wt% silica nanofiller to epoxy resin using TEM/ET. © 2012 IEEE. Reprinted, with permission, from Meichsner, C., Clark, T., Groeppel, P., Winter, B., Butz, B., and Spiecker, E. (2012). Formation of a Protective Layer During IEC (b) Test of Epoxy Resin Loaded with Silica Nanoparticles, *IEEE Trans. Dielectr. Electr. Insul.*, **19**(3), pp. 786–792.

TEM/ET is useful in determining the distance between nanofiller particles and their distribution within an observation region of several hundred of nanometer. How is the distance between nanofiller particles evaluated in wider macroscopic target regions? The method using X-rays is useful in macroscopically determining the mean distance between nanofiller particles.

When a specimen is irradiated with X-rays of a particular wavelength, the scattered X-rays show a diffraction pattern specific to each material depending on the arrangement of atoms and molecules. X-ray diffraction (XRD) is a technique of determining the distance between crystal planes, strain, and the orientation of the surface of a material from diffraction patterns. XRD can also be used to determine the mean distance between nanofiller particles in a nanocomposite.

Wide-angle X-ray diffraction (WAXD) is a technique of obtaining the diffraction images of X-rays scattered at a wide angle of $\geq 5°$ and is used to analyze Å-order structures. Figure 7.10 shows the X-ray diffraction spectra of a nanocomposite obtained by adding 5 wt% layered silicate to polyamide [9]. Unlike the spectrum of polyamide alone, a diffraction peak at 9.5° is observed in the spectrum of the nanocomposite. This finding indicates that the layered silicate in the nanocomposite is arranged periodically with a mean interval of 0.93 nm. Another diffraction peak is observed at 8.5° in the spectrum of the nanocomposite subjected to a partial discharge or plasma to deteriorate the surface. This finding indicates the possibility that the distance between nanofiller particles increases or a new structure appears inside or at the surfaces of the nanofiller particles. As explained, WAXD is used as a technique of examining the existence, separation, and agglomerates of layered silicate as well as the state of the interface.

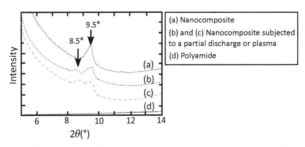

Figure 7.10 X-ray diffraction spectra of a nanocomposite obtained by adding 5 wt% layered silicate to polyamide [9].

Small-angle X-ray scattering (SAXS) is a technique of measuring X-rays scattered at a small angle of < 5° and is used to analyze the structure of a specimen 1–100 nm in size. Figure 7.11a shows the SAXS profiles of a nanocomposite consisting of colloidal silica with a particle size of 110 nm and resin. The profiles were obtained using the high-brilliance synchrotron radiation generated at SPring-8 [10]. The peak in the range of q = 0.02–0.06 nm^{-1} shifts rightward with increasing filling amount of nanofiller. Peaks of $q \geq 0.08$ nm^{-1} agree among different filling amounts, indicating that there is no aggregation of nanofiller particles. Figure 7.11b shows the distance between nanofiller particles calculated on the basis of the peak values. With increasing filling amount of nanofiller, the number density of nanofiller particles increases and the distance between nanofiller particles decreases. The results in Fig. 7.11b were observed to be similar to those obtained by electron microscopy observation.

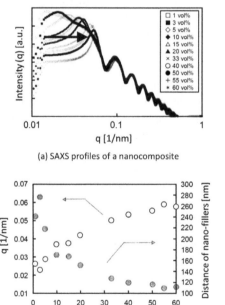

(a) SAXS profiles of a nanocomposite

(b) Distance between nano-filler particles calculated from SAXS profiles

Figure 7.11 SAXS profiles and distance between nanofiller particles of a nanocomposite consisting of colloidal silica with a particle size of 110 nm and resin [10].

7.2.4 Some Methods Are Proposed to Investigate Organic and Inorganic Bonding States

Understanding the state of chemical bonding at the interface between the polymer and the nanofiller is helpful in clarifying the interface structure of the nanocomposite. Researchers proposed a method of evaluating nanocomposites by Fourier transform infrared spectroscopy (FT-IR), which reveals the state of chemical bonding on the basis of the spectra of the transmitted light from a specimen irradiated with infrared light.

Figure 7.12 shows the FT-IR spectra of (a) epoxy resin and (b) an epoxy resin/ZnO nanocomposite obtained by adding a ZnO nanofiller to the epoxy resin. The two spectra are compared to evaluate the state of chemical bonding at the interface between the epoxy resin and the nanofiller [11]. No special surface treatment was applied to the nanofiller interface. A peak corresponding to a wavenumber of 3500 cm^{-1} is observed only in the spectrum of

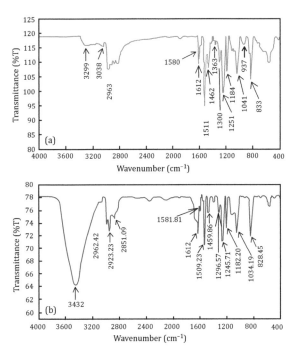

Figure 7.12 FT-IR spectra of epoxy resin and epoxy resin/ZnO nanocomposite [11]. (a) FT-IR spectra of epoxy resin. (b) FT-IR spectra of epoxy resin/ZnO nanocomposite.

Structures of Polymer/Nanofiller Interfaces

the nanocomposite. This peak indicates the existence of OH groups. The OH groups are considered to exist at the interface between the nanofiller and the polymer as shown in Fig. 7.13 rather than inside the ZnO nanofiller particles. As explained, FT-IR is a useful technique of analyzing the state of chemical bonding at interfaces and is expected to be applied to various nanocomposites such as those using nanofiller treated with a silane coupling agent.

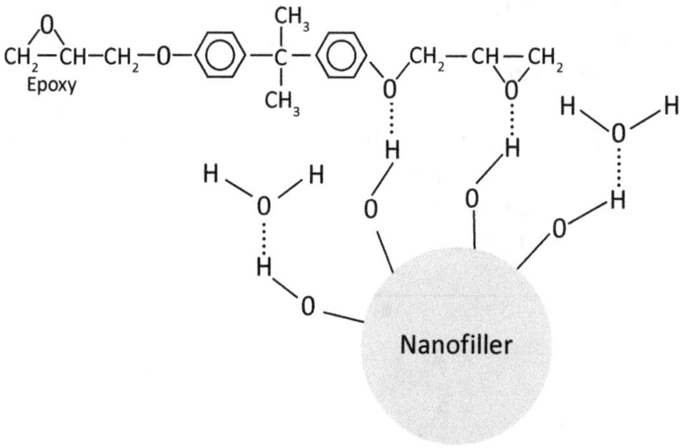

Figure 7.13 Interface model of epoxy resin/ZnO nanocomposite [11].

References

1. Ray, S. S., and Okamoto, M. (2003). Polymer/Layered Silicate Nanocomposites: A Review from Preparation to Processing, *Prog. Polym. Sci.*, **23**, pp. 1539–1641.

2. Kurimoto, M., Kai, A., Kato, K., and Okubo, H. (2008). Quantitative Analysis on the Particle Dispersibility of Epoxy/Alumina Nanocomposites, *Papers Technical Meeting on Dielectrics and Electrical Insulation, IEE Japan*, No. DEI-08-20, pp. 7–13 (in Japanese).

3. Park, J., Lee, C., Lee, J., and Kim, H. (2011). Preparation of Epoxy/ Micro- and Nano-Composites by Electric Field Dispersion Process and Its Mechanical and Electrical Properties, *IEEE Trans. Dielectr. Electr. Insul.*, **18**(3), pp. 667–674.

4. Fréchette, M. F., Larocque, R. Y., Trudeau, M., Veillette, R., Rioux, R., Pélissou, S., Besner, S., Javan, M., Cole, K., Ton That, M.-T.,

Desgagnés, D., Castellon, J., Agnel, S., Toureille, A., and Platbrood, G. (2008). Nano-Structured Polymer Microcomposites: A Distinct Class of Insulating Materials, *IEEE Trans. Dielectr. Electr. Insul.*, **15**(1), pp. 90–105.

5. Higasikoji, M., Tominaga, T., Kozako, M., Ohtsuka, S., Hikita, M., Tanaka, T., Ueta, G., and Okabe, N. (2010). Preparation of Silicone Rubber Nanocomposites and Quantitative Evaluation of Dispersion State of Nanofillers, *The Papers of Technical Meeting on Dielectrics and Electrical Insulation, IEE Japan*, No. DEI-10-065, pp. 39–44 (in Japanese).

6. Calebrese, C., Hui, L., Schadler, L. S., and Nelson, J. K. (2011). A Review on the Importance of Nanocomposite Processing to Enhance Electrical Insulation, *IEEE Trans. Dielectr. Electr. Insul.*, **18**(4), pp. 938–945.

7. Kurimoto, M., Okubo, H., Kato, K., Hanai, M., Hoshina, Y., Takei, M., and Hayakawa, N. (2010). Permittivity Characteristics of Epoxy/ Alumina Nanocomposite with High Particle Dispersibility by Combining Ultrasonic Wave and Centrifugal Force, *IEEE Trans. Dielectr. Electr. Insul.*, **17**(3), pp. 662–670.

8. Meichsner, C., Clark, T., Groeppel, P., Winter, B., Butz, B., and Spiecker, E. (2012). Formation of a Protective Layer During IEC(b) Test of Epoxy Resin Loaded with Silica Nanoparticles, *IEEE Trans. Dielectr. Electr. Insul.*, **19**(3), pp. 786–792.

9. Tanaka, T., Kozako, M., Fuse N., and Ohki, Y. (2005) Proposal of a Multi-Core Model for Polymer Nanocomposite Dielectrics, *IEEE Trans. Dielectr. Electr. Insul.*, **12**(4), pp. 669–681.

10. Senoo, M., Takeuchi, K., Oka, A., Shimonabe, Y., Kuwamoto, S., Urushibara, Y., Matsui, S., and Nakamae, M. (2009). *Proc. 59th the Network Polymer Symposium Japan*, pp. 29–32 (in Japanese).

11. Singha, S., and Thomas, M. J. (2009). Influence of Filler Loading on Dielectric Properties of Epoxy-ZnO Nanocomposites, *IEEE Trans. Dielectr. Electr. Insul.*, **16**(2), pp. 531–542.

Chapter 8

Computer Simulation Methods to Visualize Nanofillers in Polymers: Toward Clarification of Mechanisms to Improve Performance of Nanocomposites

Masahiro Kozako[a] and Atsushi Otake[b]

[a]*Kyushu Institute of Technology*
[b]*Hitachi, Ltd.*

8.1 Non-Empirical (ab initio) Molecular Orbital Method

Masahiro Kozako

Department of Electrical and Electronic Engineering, Kyushu Institute of Technology, Kitakyushu-shi, Fukuoka 804-8550, Japan

kozako@ele.kyutech.ac.jp

It is necessary to consider electronic states based on quantum mechanics in order to interpret chemical and physical phenomena of polymers theoretically from their electronic properties. Several techniques have been developed for calculating the electronic states of each of the molecules and atoms in polymers in detail. What physical properties of polymer nanocomposites can be

---

Advanced Nanodielectrics: Fundamentals and Applications
Edited by Toshikatsu Tanaka and Takahiro Imai
Copyright © 2017 Pan Stanford Publishing Pte. Ltd.
ISBN 978-981-4745-02-4 (Hardcover), 978-1-315-23074-0 (eBook)
www.panstanford.com

elucidated with the use of a computational technique based on ab initio? Such a technique has been applied to polymers as dielectric materials since around 2000, by which basic parameters related to electrical conductivity and dielectric breakdown have been estimated, including band structures, trap levels, electron affinity, and carrier mobility.

8.1.1 Computer Simulation of Nanocomposites Has Begun

There has been remarkable evolution in the field of computational chemistry using computer simulation techniques in recent years. These techniques have been applied in the study and development of chemicals and semiconductor materials. Simulation techniques useful in this field can be classified as follows in descending order from the smallest unit of the handling system:

(i) First-principle (ab initio) molecular orbital (MO) method
(ii) Molecular dynamics (MD) method
(iii) Monte Carlo method

Methods (i) and (ii) are addressed here. The ab initio MO methods include the method starting from the one-electron approximation of electronic states and the density functional theory (DFT), but the easier ab initio DFT MO method is often used these days. Because the MD method can handle multi-molecular systems with less computer resources, it is applied to complex systems that contain many inter-molecular interactions. Such systems have been used extensively in the field of dielectric insulating materials since around 2000.

Although simulation techniques for nanocomposites have only recently been initiated, some are currently being studied, including structural analysis of the nanofiller/polymer interface and analysis of dispersion behavior of nanofillers. These techniques will be important in ensuring the qualitative and quantitative derivation of not only mechanical and thermal properties but also dielectric and insulation properties of polymer nanocomposites, and to enable manufacture and development of new excellent nanocomposite materials in the future. Therefore, further development of computer simulation techniques is expected hereafter.

8.1.2 What Is ab initio Calculation Like?

Computer simulation methods are classified as shown in Fig. 8.1 in accordance with the size and time scale for the target phenomenon. They include (i) quantum mechanics dealing with electronic states, (ii) molecular dynamics and Monte Carlo simulation dealing with the collective motion of atoms and molecules, (iii) finite element method and statistical thermodynamics calculation dealing with bulk materials, and (iv) phase-field method dealing with a mesoscale that connects between micro and the macro scaled phenomena. This paper focuses on the molecular orbital method, which is widely used as a calculation method based on quantum mechanics.

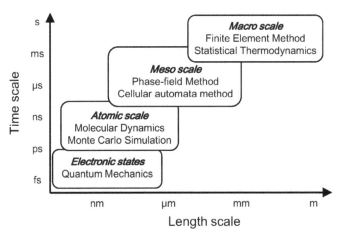

Figure 8.1 Relationship between time scale and length scale of each calculation method.

Molecular orbital (MO) theory is the basis of almost all modern quantum mechanical calculations of the electronic structure of molecules. This calculation method is divided broadly into two types: the semi-empirical molecular orbital method and the non-empirical molecular orbital method (or the ab initio MO method). The former introduces empirical parameters into the calculation process for simplifying calculations. The latter is a solution without approximation from the beginning. "ab initio" is a Latin term meaning "from the beginning."

The ab initio MO method is described here. The ab initio MO method has been used on the basis of the Hartree–Fock approximation (one-electron wave function) or the post-Hartree–Fock approximation (incorporating electron correlation). However, a simpler method has been developed by incorporating the DFT into the calculation of electronic states. The ab initio DFT method has the feature in which a precise result with less computation time can be presented in consideration of a certain degree of electron correlation by representing electron correlation as a function of electron density. Therefore, the ab initio DFT method has often been used in recent years.

8.1.3 What Can Be Found When Using the ab initio Calculation?

8.1.3.1 Interfaces account for enormous rate in nanocomposites

An example of an application of the ab initio localized dielectric theory is introduced here, which demonstrates a change in the dielectric constant of the interface between nanofiller and polymer in a polymer nanocomposite [1]. Figure 8.2 shows the static and optical permittivity profile across a polyethylene–silica (SiO_2) interface as a function of position along the direction normal to the interface. The terms "static" and "optical" in the figure signify a low-frequency component and a high-frequency component, respectively. Since the interface is coupled with different materials, interfacial polarization clearly appears and an enhancement in the dielectric constant is observed. When treating the interfacial bonds using a silane coupling agent, more realistic simulation can be performed. It can be found from the localized dielectric phenomenology that a change in the dielectric constant at the interface between nanofiller and polymer has a correlation with a chemical bonding state at the interface. It would be possible in the future to estimate effects of localized physical properties at the interface on the dielectric constant of the bulk by extending such calculations on an enormous scale.

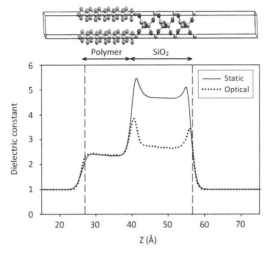

Figure 8.2 Top: Atomic model of SiO$_2$-polymer, with Si shown in gold, O in red, C in yellow, and H in green. Bottom: Static (solid) and optical (dotted) dielectric constant of C$_{12}$H$_{25}$-SiO$_2$ stack as a function of position z normal to the interface. © 2008 IEEE. Reprinted, with permission, from Shi, N., and Ramprasad, R. (2008). Local Properties at Interfaces in Nanodielectrics: An ab initio Computational Study, *IEEE Trans. Dielectr. Electr. Insul.*, **15**, pp. 170–177.

8.1.3.2 Electron trap sites can be estimated

There is a case in which electron trap sites at the interface between nanofiller and polymer in polymer nanocomposite are considered on the basis of ab initio calculation based on the layer-decomposed density of states [1]. Figure 8.3 shows atomic structures of the silica (SiO$_2$)/vinylsilanediol/polyvinylidene fluoride (PVDF) interface and the results of calculating the layer-decomposed density of states at the atomic level. The conduction band is shown in blue and the valence band in black. The zero of energy is Fermi energy. The lower and upper circled states represent occupied and unoccupied interfacial defect states, respectively. Since these defect states are close to the band edges, they could serve as shallow traps or "hopping sites" for electrons or holes. Hence, the appropriate choice of initiator is critical in determining charge transport along or perpendicular to the interface.

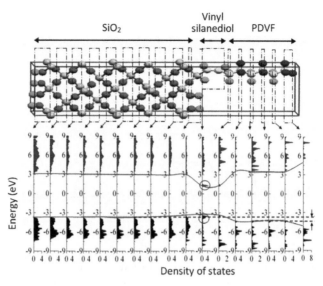

Figure 8.3 The conduction band is shown in blue, and the valence band in black. The zero of energy is the Fermi energy. Lower and upper circled states represent occupied and unoccupied interfacial defect states, respectively. © 2008 IEEE. Reprinted, with permission, from Shi, N., and Ramprasad, R. (2008). Local Properties at Interfaces in Nanodielectrics: An ab initio Computational Study, *IEEE Trans. Dielectr. Electr. Insul.*, **15**, pp. 170–177.

On the other hand, the depth of electron traps is considered on the basis of electron affinity and electrostatic potential distribution in various polymers such as polar polymers and non-polar polymers [2]. Some experimental results show that decomposed residues of cross-linking agents in cross-linked polyethylene have a great influence on accumulated charges. Therefore the ab initio calculation was performed to verify it. It was thus revealed that acetophenone as a decomposition residue of the cross-linking agent may form a trap site [3]. There have also been investigations on the generation and movement of charges in a high electric field in the vicinity of water molecules in a polyimide in order to verify moisture effects of accumulated charges in the polyimide [4]. In these studies, it has been argued that quantum chemical calculations based on the MO method could be very effective.

It will be necessary to make comparative examinations of these analyses and experiments through further data accumulation in the future, which should be useful for elucidating mechanisms such as electron conduction and space charge formation in a polymer nanocomposite and thus promote development of innovative multifunctional polymer nanocomposite materials.

8.1.4 Application to Large-Scale Systems Is Also Attempted

The MO method is usually less suitable for calculating large-scale systems. Since it is a method for diagonalizing each integral element, a calculation amount up to as much as the fourth power of the number of atoms or tracks is required in the ab initio MO method. The amount of calculation becomes enormous with increase in the number of atoms, and it becomes a problem that cannot be solved even with a supercomputer. As described in the next section, a combination of the MO method with molecular dynamics has been applied in some cases, which is also useful. In the meantime, examination of large-scale systems has continued, one of which is the fragment molecular orbital method (FMO). While the FMO method has been used to deal with the interaction of huge organic molecules, further improvements of the calculation process itself are underway to ensure higher accuracy and reproducibility.

References

1. Shi, N., and Ramprasad, R. (2008). Local Properties at Interfaces in Nanodielectrics: An ab initio Computational Study, *IEEE Trans. Dielectr. Electr. Insul.*, **15**, pp. 170–177.

2. Hayase, Y., Tahara, M., Takada, T., Tanaka, Y., and Yoshida, M. (2009). Relationship between Electric Potential Distribution and Trap Depth in Polymeric Materials, *IEEJ Trans. Fundamentals Mater.*, **129**(7), pp. 455–462 (in Japanese).

3. Takada, T., Hayase, Miyake, H., Tanaka, Y., and Yoshida, M. (2012). Study on Electric Charge Trapping in Cross-linking Polyethylene and Byproducts by Using Molecular Orbital Calculation, *IEEJ Trans. Fundamentals Mater.*, **132**(2), pp. 129–135 (in Japanese).

4. Takada, T., Ishii, T., Komiyama, Y., Miyake, H., and Tanaka, Y. (2013). Discussion on Hetero Charge Accumulation in Polyimide Film Containing Water by Quantum Chemical Calculation, *IEEJ Trans. Fundamentals Mater.*, **133**(5), pp. 313–321 (in Japanese).

8.2 Simulation of Performance of Nanocomposites by Coarse-Grained Molecular Dynamics

Atsushi Otake

Hitachi, Ltd., Hitachi-shi, Ibaraki 319-1292, Japan

atsushi.otake.bg@hitachi.com

In this section, the molecular dynamics simulation is explained. Among various molecular simulations, the molecular dynamics simulation has the feature of low computational load and can treat relatively large analytical systems. In particular, coarse-grained molecular dynamics can treat micron-order systems, which are larger than those treated in conventional molecular dynamics simulations, and can simulate the behavior of particles with sizes that can be observed by scanning electron microscopy (SEM).

8.2.1 What Is Molecular Dynamics?

8.2.1.1 Overall concept of molecular dynamics

Molecular dynamics is a computational chemistry simulation technique for clarifying the properties and behavior of groups of atoms and molecules. This technique was studied along with molecular orbital theory, a typical method of quantum chemical calculation. In molecular dynamics (including classical molecular dynamics but excluding first-principles molecular dynamics), large systems can be treated with a small number of computational resources because the equations treated in molecular dynamics follow those treated in classical mechanics.

Molecular dynamics was proposed in the 1950s. As the performance of computers has improved, the number of particles that can be treated in molecular dynamics has increased, and the method of determining the force fields that act between particles has improved. The shape of the force fields is determined from measurements obtained from spectroscopy or from thermodynamics data. In addition, force fields are also derived using molecular orbital theory; in concrete terms, the total energy of a system with different arrangements of atoms and molecules (corresponding to the stabilization energy, or the heat of formation, of the groups of atoms and molecules in the system) is calculated, and then the parameters of the force fields as a function of the coordinates of the atoms are determined. How to calculate the parameters is explained in the following using Fig. 8.4. A function that can fit the curve of total energy (the potential curve) determined by the molecular orbital analysis of interatomic distances is desirable; however, there are no functions that can fit every target curve. Therefore, a function that can fit the potential curve determined by molecular orbital calculations over a critical region is selected. Fitting is performed using the nonlinear least-squares method to determine the parameters of the function f. The parameters in Fig. 8.4 are the spring constant (force constant) between two atoms and the equilibrium interatomic distance. In general, the parameters of f include the equilibrium interatomic distance, equilibrium bond angle, and the interatomic vibrational and rotational force constants.

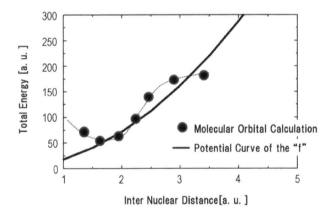

Figure 8.4   Determination of parameters of f by molecular orbital simulation.

The potential curve of f has a minimum value at the interatomic distance at which a stable bond is formed. Because of this, the total energy of the entire system formed by multiple atoms tends to become minimum. In accordance with the calculation time step, the molecules and atoms move like balls to stabilize the entire system.

Table 8.1 shows a brief summary of molecular dynamics methods. As of 2015, both commercial and free software packages of molecular dynamics simulations with different features are available. Simulation targets depend on the force fields and simulation method adopted in each software package. For example, the target applications include (i) biotechnology, (ii) organic polymers, and (iii) inorganic compounds. For groups of organic polymers and those containing a nanosized filler, the use of general molecular dynamics simulations results in a huge number of calculations, denoting that the simulation is practically impossible, because the movement of all of the many interacting atoms is traced in the simulations. A solution to this problem is coarse-grained molecular dynamics.

Table 8.1 Rough classification of molecular dynamics

Method	Determination of force-field parameters	Features
Classical molecular dynamics	Experimental data (e.g., spectroscopy and thermodynamics data)	The number of types of compounds that can be simulated well is limited. Simple and low-cost simulation is possible.
Semiclassical molecular dynamics	Experimental data, molecular orbital simulation	Accurate simulation is possible when the force-field parameters are determined by highly approximated molecular orbital simulation.
First-principles molecular dynamics	Usually, force fields are unnecessary	The interatomic force is quantum-calculated for each time step to determine the position of the atoms. A large number of calculations are required.

8.2.1.2 Advantages of coarse-grained molecular dynamics

Two approaches are available for predicting the behavior of fine particles in resin before hardening and the stress of thermosetting resin: (i) the continuum model, in which liquids and solids are treated as continuums, similarly to fluid dynamics and continuum dynamics, and (ii) molecular dynamics, in which all atoms constituting both solutions and resin are considered with their interactions.

The continuum model is effective for optimizing the structure of stirrers and predicting the sites of stress concentration that depend on the shape of thermosetting resin. However, physical properties, such as the viscosity of the solutions and Young's modulus of the thermosetting resin, must be known before simulation. Therefore, this model is not suitable for predicting the physical properties of new materials that require trial-and-error approaches. In contrast, the molecular dynamics model is considered suitable for the development of new materials because the physical properties of targets are estimated from the interactions between their atoms and molecules. However, analytical systems of at least several tens of nanometers in size (that is, at least 10^8 atoms are taken into consideration) are required to simulate the effect of the nanosized filler. Moreover, the actual time taken for the stirring of fine particles in solution and the rupture of a thermosetting resin is microseconds to several seconds or longer, requiring 10^6–10^{12} or more calculation cycles for the time step of molecular dynamics (picoseconds or shorter, depending on the frequency of atomic vibration), in which all atoms are taken into consideration. Thus, it is difficult to simulate the stirring of fine particles in solution and the stress characteristics of a thermosetting resin while considering all atoms.

Meanwhile, the coarse-grained molecular dynamics technique has been developed for simulating the behavior of molecular sets in solvent and has been improved and put into practical use [1, 2]. In this method, a set of molecules is regarded as a single particle ("coarse-grained"), as shown in Fig. 8.5, enabling the simulation of large analytical systems. Because the properties of molecules can be reflected in the interactions between particles, the simulation technique can also be used to develop new materials. The interactions between particles can be determined by the ab initio molecular orbital method.

In this section, previous reports and presentations on the physical properties of a nanocomposite resin simulated by a platform called OCTA [3, 4] for designing high functional materials and a coarse-grained molecular dynamics program by Nagoya Cooperation (COGNAC) [4], a component of OCTA, are introduced.

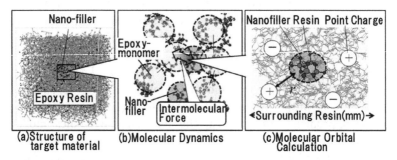

Figure 8.5 Conceptual schematics of coarse-grained molecular dynamics based on molecular orbital theory [2].

8.2.1.3 Calculation method

In molecular dynamics, the temporal development of the position of particles is solved by following classical mechanics. In coarse-grained molecular dynamics, sets of particles, instead of single atoms, are considered. The basic equation is given by

$$m_i \frac{d^2 \vec{r_i}}{dt^2} = \sum_{i \neq j} \vec{f}_{ji} \qquad (8.1)$$

Here, the subscripts i and j are the particle set numbers, m is the mass of a particle, r_i is the coordinate of the particle set i, and f is the force (force field) acting between the particle sets i and j. The analytical system suitable for simulation is determined depending on the expression of f. The reproducibility can be increased by using an f appropriate for the phenomenon, although f may be given as a function of a simple spring model.

When a small set of nanofillers, that is, a set of molecules, is dispersed in a solvent, dissipative particle dynamics (DPD), as described in Refs. [6, 7], can be used for simulation.

$$f_{ji} = \sum_{i \neq j} F_{ij}^D + F_{ij}^R + F_{ij}^C + F_{ij}^S \qquad (8.2)$$

The terms F^D and F^R are called the dissipative and random forces, respectively, and are the terms in which the averaged effect of molecular vibrations ignored by coarse-graining is considered. For the distance between the particles r_{ij}, F^D and F^R are given by

$$F_{ij}^D = -\gamma \omega^D(r_{ij}) \cdot (\vec{r}_{ij} \cdot \vec{v}_{ij}) \frac{\vec{r}_{ij}}{r_{ij}} \tag{8.3}$$

$$F_{ij}^R = \sigma \omega^R(r_{ij}) \cdot \theta_{ij} r_{ij} \tag{8.4}$$

The term F^D in Eq. (8.3) is the friction term that decreases the relative velocity between the particles v_{ij} and represents the dissipation of energy. In Eq. (8.4), θ_{ij} is a factor randomly generated with respect to time $(-1 < \theta < 1)$ and gives thermal energy to the particle sets i and j. The energy lost by dissipation and the thermal energy acquired by random thermal movement are the terms attributable to the exchange of molecular vibration energy between the particles. The terms ω^D and ω^R are the distances affected by F^D and F^R, respectively. The term ω^R is given by

$$\omega^R(r_{ij}) = \begin{cases} \left(1 - \dfrac{r_{ij}}{r_c}\right) & (r_{ij} < r_c) \\ 0 & (r_{ij} > r_c) \end{cases} \tag{8.5}$$

Because F^D and F^R are not conservative forces, thermal equilibrium holds between the energy lost by dissipation and that acquired by random thermal movement as follows [7]:

$$\omega^D(r_{ij}) = \{\omega^R(r_{ij})\}^2$$

$$\sigma^2 = 2\gamma k_B T \tag{8.6}$$

The term F^C in eq. (8.2) is a repulsive force between particles introduced to prevent the particles from excessively approaching each other and is given by the following model equation.

$$F_{ij}^C = \begin{cases} a_{ij}\left(1 - \dfrac{r_{ij}}{r_c}\right)\vec{r}_{ij} & (r_{ij} < r_c) \\ 0 & (r_{ij} > r_c) \end{cases} \tag{8.7}$$

Here, r_c is the range of distances affected by F^C and should be set to several nanometers, although its optimal value depends on the system. The value a_{ij} represents the magnitude of F^C. In Ref. [8], $a_{ij} = 25 \times k_B T$ is proposed (k_B, the Boltzmann constant; T, temperature).

The term F^S represents the force of the chemical bond between epoxy monomers that form polymers and is given as the following spring-type force.

$$F_{ij}^S = -C \cdot \vec{r}_{ij} \tag{8.8}$$

Here, C is a force constant. The above equations are an example of rough modeling, and force fields should be modeled in accordance with each analytical system.

8.2.1.4 Accurate determination of force fields for large analytical systems

The force field used in coarse-grained molecular dynamics is a function of the force acting between sets of particles. As mentioned above, this function can be determined following molecular orbital theory. This theory originated from the discovery of quantum mechanics and is used to approximately solve the time-independent Schrödinger equation. Because of its high computational load, the mainstream of molecular orbital theory consisted of semi-empirical approaches with the problem of low accuracy until the emergence of supercomputers.

In the 1980s, ab initio molecular orbital simulation became generally available, and many software applications for Hartree–Fock approximation using Gaussian basis functions were developed. The reason why the Gaussian basis functions were adopted is explained in the following. In general, the existence probability of electrons around an atom forms a Slater-type distribution $(\exp[-\alpha \cdot x])$. Because analytically solving the integral generated during the calculation process is difficult, numerical integration with a high computational load is used. When a Gaussian function $(\exp[-\beta \cdot x^2])$ is linearly combined with the existence probability to form an almost Slater-type distribution, analytical integration becomes possible and the number of calculations is reduced. For general molecules, the calculation accuracy is increased by using a sufficient number of Gaussian functions, but significant errors may be generated for some molecules. In addition, the Hartree–Fock method has a drawback in that the many-body interaction of electrons (electron correlation) is not taken into consideration.

Currently available molecular orbital simulations are based on a method with higher accuracy that considers electron correlation [the many-body perturbation method, the configuration interaction (CI) method] or density functional theory with a high accuracy per number of calculations. Molecular orbital simulations consume computational resources particularly for systems with many atoms and electrons. To solve this problem, rapid analysis methods, such as a method that treats distant atoms by approximation, have been developed and used for the calculation of potential energy.

8.2.2 Examples of Practical Use of Coarse-Grained Molecular Dynamics

In this section, examples of applying the previously outlined coarse-grained molecular dynamics method to the simulation of the properties of nanocomposite resin are introduced. There is still room for the improvement of computational chemistry approaches, which are expected to play a key role in the development of nanocomposites in the future.

8.2.2.1 How is a nanofiller dispersed in resin?

The properties of a nanocomposite resin generally depend on the dispersibility of the nanofiller. Therefore, it is important to simulate the degree of dispersibility of the nanofiller that improves the properties of the nanocomposite resin. In many cases, the higher the dispersibility, the more improved the resin properties. However, other cases may arise. In the following, such a dispersion state is examined using the simulation results obtained by coarse-grained molecular dynamics and from experimental results.

Figure 8.6 shows schematics of two types of silica nanofiller used in Ref. [5]. One is ordinary silica, which has a silanol group exposed on its surface and is highly hydrophilic. The silanol group has a high affinity for polar solvents because oxygen and hydrogen atoms of O–H bonds are negatively and positively charged, respectively. The hydrophilic silica interacts with the charges on the epoxy resin, which is generally used and has high polarity. The other silica nanofiller rarely interacts with the epoxy resin because almost all O–H groups (the source of polarity) on the surface are methylated.

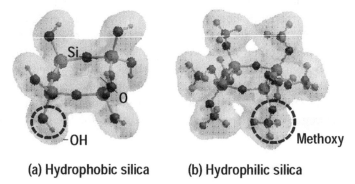

(a) Hydrophobic silica   (b) Hydrophilic silica

Figure 8.6 Electron density distribution of silica nanofiller in epoxy resin [5].

In Ref. [5], the dispersion states of the hydrophilic and hydrophobic silica nanofillers in an epoxy resin are simulated by coarse-grained molecular dynamics. The simulation results are compared with the experimental results, as shown in Fig. 8.7.

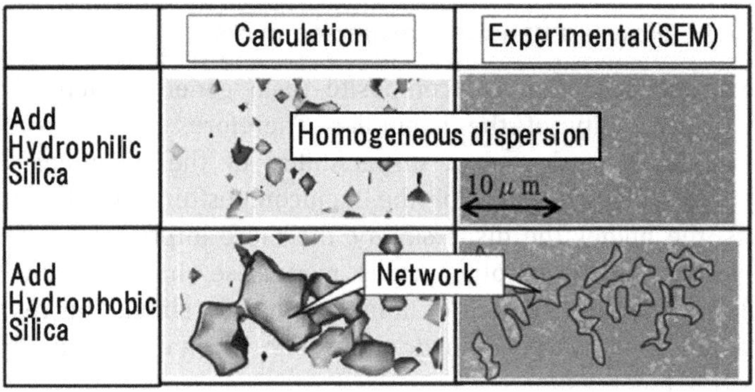

Figure 8.7 Simulation and experimental results for dispersion states of two types of silica nanofiller [5].

The dispersion states of the two types of silica nanofiller in an epoxy resin are simulated well by coarse-grained molecular dynamics. The results indicate that a network structure is formed inside the epoxy resin when a hydrophobic silica nanofiller is dispersed. This structure is considered to strengthen the resin, and the strength of the resin is calculated in Ref. [5]. Details of the calculated results are explained in the next section.

8.2.2.2 Type of nanofiller and mechanical properties of resin

Figure 8.8 shows the calculated strength and stress characteristics of an epoxy resin with two types of silica nanofiller dispersed. Both simulated and experimental results indicate that the resin with the hydrophobic silica nanofiller is stronger than that with the hydrophilic silica nanofiller. Thus, uniform dispersion does not always result in good material characteristics.

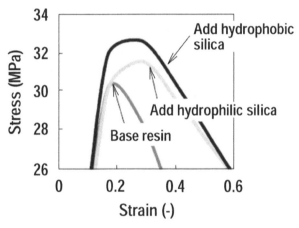

Figure 8.8   Effect of two types of silica nanofiller on improving strength and stress characteristics of resin [5].

The technique for analyzing the mechanical strength of the nanofiller was also applied to elastomers. Figure 8.9 shows the simulated and experimental stress characteristics of an elastomer. The tendency of the simulation results is in agreement with that of the experimental results, clarifying the effect of the filler size on the toughness of the epoxy resin. This indicates the possibility of a first-principles-based and semi-quantitative simulation of the relationship between the dispersion state of the nanofiller in an epoxy resin and the strength of the resin.

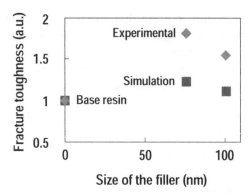

Figure 8.9 Relationship between size of fine elastomer particles and toughness of resin [9].

8.2.2.3 What occurs in the vicinity of a nanofiller?

In Ref. [11], the interfacial affinity between the nanofiller and polymers was simulated by COGNAC. Figure 8.10 shows the model system used in the simulation. Linear molecules (number of segments, 40) are located between two walls. A strong attractive force is assumed to act between the nanoparticles forming the walls and the linear molecules. For the model system, the mean square displacement of the linear molecules (segments) is also calculated for different temperatures (T). The results are shown in Fig. 8.11.

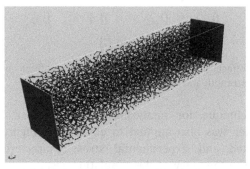

Figure 8.10 Model interface between nanofiller and polymers [11].

The results indicate that with increasing T, the constrained layer of the linear molecules shrinks and the travel distance of the molecules increases. Thus, the changes in the position and width of the constrained layer of the linear molecules in the vicinity of the walls for different temperatures are successfully simulated.

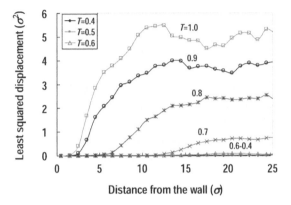

Figure 8.11 Relationship between displacement of linear molecules and temperature [11].

As described previously, the semi-empirical or ab initio analyses of the dispersion of a nanofiller in a resin and the properties of the resin have been gradually realized by coarse-grained molecular dynamics and a combination of coarse-grained molecular dynamics and other simulation techniques. Research on coarse-grained molecular dynamics has just started, but this method has attracted attention as a simulation technique that is effective for predicting the effect of the nanofiller, which is currently studied by approaches based on guess work.

References

1. Sawa, F., Imai, T., Ozaki, T., Shimizu, T., and Tanaka, T. (2007). Molecular Dynamics Simulation of Characteristics of Polymer Matrices in Nanocomposites, *Annual Rept. IEEE CEIDP*, pp. 264–266.
2. Sano, A., Ohtake, A., Kobayashi, K., and Matsumoto, H. (2011). Development of Simulation Techniques for Mechanical Strength of Nanocomposite Insulating Materials, *Proc. IEEJ ISEIM*, No. D2, pp. 77–79.

3. Website of OCTA, a Platform for Designing High Functional Materials: http://octa.jp/index_jp.html (in Japanese).

4. Aoyagi, T., Sawa, F., Shoji, T., Fukunaga, T., Takimoto, J., and Doi, M. (2002). A General-Purpose Coarse-Grained Molecular Dynamics Program, *Comput. Phys. Commun.*, **145**, pp. 267–279.

5. Sano, A., Ohtake, A., Kobayashi, K., and Matsumoto, H. (2013). Development of Simulation Techniques of Mechanical Strength of Nano-Composite Insulating Materials, *IEEJ Trans. Fundamentals Mater.*, **133**, pp. 81–84.

6. Groot, R. D., and Warren, P. B. (1997). Dissipative Particle Dynamics: Bridging the Gap between Atomistic and Mesoscopic Simulation, *J. Chem. Phys.*, **107**, pp. 4423–4435.

7. Scocchi, G., Posocco, P., Fermeglia, M., and Pricl, S. (2007). Polymer-Clay Nanocomposites: A Multiscale Molecular Modeling Approach, *J. Phys. Chem. B*, **111**, pp. 2143–2151.

8. Barnes, J., and Hut, P. (1986). A Hierarchical Force Calculation Algorithm, *Nature*, **324**, pp. 446–449.

9. Kato, T., Sano, A., Matsumoto, H., Ohtake, A., and Kobayashi, K. (2012). Development of Analysis Technique for Mechanical Strength of Nanocomposite Insulating Resin: Application to Silica/Epoxy and Rubber/Epoxy Composites, *Proc. IEEJ Annu. Conf. Fundamentals Mater.*, No. XVI-3, p. 338 (in Japanese).

10. Sawa, F., Imai, T., Ozaki, T., Shimizu, T., and Tanaka, T. (2009). Coarse Grained Molecular Dynamics Simulation of Thermosetting Resins in Nanocomposite, *Proc. IEEE ICPADM*, pp. 853–856.

11. Sawa, F., Imai, T., and Sato, J. (2010). Simulation Technique for Development of Insulating Materials, *Proc. 41st IEEJ ISEIM*, No. SS-10, pp. 265–266 (in Japanese).

Chapter 9

Epilogue: Environmental Concerns and Future Prospects

Takahiro Imai

Power and Industrial Systems R&D Center, Toshiba Corporation, Fuchu-shi, Tokyo 183-8511, Japan

takahiro2.imai@toshiba.co.jp

This book, entitled *Advanced Nanodielectrics*, describes applications and fundamentals of nanocomposite insulation materials. There are indeed concerns about the harmful effects of nanomaterials on the human body and the environment, but nanocomposite insulation materials have superior properties to conventional materials. In this chapter, the first section provides the risk assessment of nanofillers, which are typical nanomaterials, and gives guidelines concerning nanofiller handling. The second section proposes future prospects in the power and industrial sector based on nanocomposite insulation materials and considers the role of nanocomposite insulation materials and their contribution to a sustainable society, which is one of the issues in the twenty-first century. This section also encourages interdisciplinary cooperation between electrical, material, chemical, physical, biological, and computer simulation engineering to accelerate the nanocomposite research in the industry and academia.

Advanced Nanodielectrics: Fundamentals and Applications
Edited by Toshikatsu Tanaka and Takahiro Imai
Copyright © 2017 Pan Stanford Publishing Pte. Ltd.
ISBN 978-981-4745-02-4 (Hardcover), 978-1-315-23074-0 (eBook)
www.panstanford.com

9.1 Awareness in Dealing with Nanofillers

9.1.1 Effects of Nanofillers on Human Body and Environment Are a Matter of Concern

As described in the previous chapters, nanocomposite insulation materials that have small amounts of homogenous dispersed inorganic fillers of nanometer dimensions (nanofiller) in a polymer matrix show insulation properties far superior to those of the original polymer, which one of their attractive characteristics. A nanofiller has a much larger surface than a micro-filler. For the same filler content, the strong interaction between nanofiller surfaces and polymer seems to improve the properties of the nanocomposite compared to the interaction attributed to microfillers.

However, there are concerns about harmful effects of nanomaterials on the human body and the environment because they are extremely small. Generally, the assessment of risks (1) to worker, (2) via environment, (3) via product, and (4) of accident is necessary, as shown in Fig. 9.1 [1]. Since most nanocomposites are still in the research and development stage, "risk to worker" should be the primary concern.

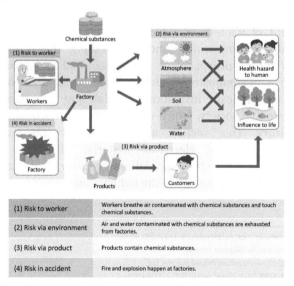

Figure 9.1 Risks attributed to chemical substances [1].

In nanofiller handling during nanocomposite research, processes in which nanofillers may be touched are shown in Fig. 9.2. In particular, workers (researchers and engineers) are more likely to inhale or touch nanofillers in the fabrication process. Therefore, both handling of the nanofiller inside a local exhaust ventilation system and wearing of appropriate protective equipment will effectively reduce the risk.

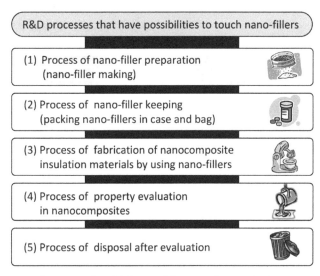

Figure 9.2 Processes in which nanofillers may be touched.

9.1.2 Risk Evaluation of Nanofillers Is in Progress

Generally, the chemical substance risk is expressed as a product of "hazardous property" and "amount of exposure" as shown in Fig. 9.3 [2]. No observed adverse effect level (NOAEL) is estimated from the laboratory animal testing, and the actual intake is measured/estimated as amount of exposure. Risk evaluation is conducted by comparison of NOAEL and actual intake. In the chemical substance risk evaluation, when the exposure and intake amount are lower than NOAEL, the incidence rate of the hazardous influence is very low, as shown in Fig. 9.4 [3]. However, when the exposure and intake amount exceed NOAEL, the incidence rate of the hazardous influence increases with increasing exposure and intake amount. In other words, if the exposure and intake amount of a chemical substance with high hazard property is lower than NOAEL, the hazard to the human body is inactive. However, if the exposure and intake amount of a chemical substance with low hazardous property is higher than NOAEL, the hazard to the human body is active. For existing chemical substances, the (Material) Safety Data Sheets ((M) SDS) are documented based on the risk evaluation. Appropriate risk management is needed according to exposure prevention, protection, handling, and storage, as indicated by the SDS.

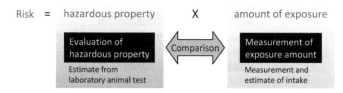

Figure 9.3 Concept for chemical substance risk evaluation [2].

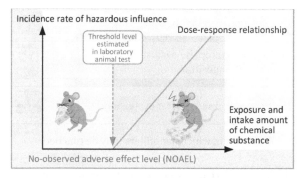

Figure 9.4 Incidence rate of hazardous influence in risk evaluation [3].

Unlike the risk evaluation of common chemical substances, the risk evaluation of nanomaterials needs sample preparation and characterization [4]. This is important and essential before the evaluation of hazardous property and measurement of exposure amount, as shown in Fig. 9.5. For example, nanofillers tend to aggregate due to their van der Waals force. The nanofillers exist as aggregate and agglomerate in the prepared sample. Therefore, the dispersion state and hazardous property of nanofillers depend on the method of sample preparation. An understanding of nanofiller characteristics such as shape, dispersion state, diameter, and surface area is necessary for the accurate risk evaluation. Moreover, in nanofillers for industrial use, impure substances (contamination) as well as nanofillers have an effect on the hazardous property. In addition, the surfaces of nanofillers are often modified chemically to improve their affinity to solvent and polymer. For the same nanofiller, the hazardous property is dependent on the dispersion state, surface modification, and contamination. Thus, sample preparation and characterization are of primary importance in risk evaluation.

The National Institute of Advanced Industrial Science and Technology (AIST) conducted a "risk assessment of manufactured nanomaterials" project supported by the New Energy and Industrial Technology Development Organization (NEDO) from June 2006 to February 2011 as typical risk evaluation of nanofillers in Japan. Risk assessment of typical nanofillers such as titania (TiO$_2$), fullerene (C60), and carbon nanotube (CNT) was conducted based on the framework of risk evaluation and management for industrial nanomaterials, as shown in Fig. 9.5 [2]. In addition to the assessment results, sample preparation method, the measurement procedures for the evaluation of the hazardous property, exhaust, and exposure were published [5]. Moreover, risk assessments have been conducted in other countries, including the United States, and assessments vary depending on the evaluators [6].

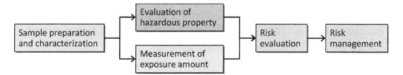

Figure 9.5 Framework of risk evaluation and management for industrial nanomaterials [2].

9.1.3 Guidelines for Dealing with Nanofillers

Recently, PM 2.5 has gained attention with respect to air pollution in China. PM 2.5 means particulate matter (PM) smaller than 2.5 μm in diameter. Suspended particulate matter (SPM) smaller than 10 μm in diameter is regulated according to environmental standards. However, there are concerns about the adverse effect of PM 2.5 on the circulatory organ system as well as the respiratory organ system because it is smaller than SPM.

Epilogue

A nanofiller is much smaller than PM 2.5. The information required to ensure the safety of the human body and the environment is lacking, although it is accumulating from risk evaluation of nanofillers. However, nanofillers have lots of specific advantages and have attracted a lot of attention in various industries and academia. The insulation properties of nano-composite materials described in this book are improved by taking advantage of nanofillers. Nanofillers have negative aspects and there are concerns about their effect on the human body and the environment. However, they have lots of positive aspects enabling the fabrication of innovative materials with novel functions.

Table 9.1 Reports and guidelines regarding nanomaterial handling in Japan

Organization/ institute	Published year	Report/guideline
Labour Standard Bureau, Ministry of Health, Labour and Welfare	November 26, 2008	Report of Review Panel Meetings on Preventive Measures for Worker Exposure to Chemical Substances Posing Unknown Risks to Human Health.
	March 31, 2009	Notification on Precautionary Measures for Prevention of Exposure, etc., to Nanomaterials.
Ministry of Environment	March 10, 2009	Guidelines for Preventing the Environmental Impact of Manufactured Nanomaterials.
AIST (National Institute of Advanced Industrial Science and Technology)	August 11, 2011	Risk Assessment of Manufactured Nanomaterials. "Approaches—Overview of Approaches and Results-" "-Fullerene (C_{60})-" "-Carbon Nanotube (CNT)-" "-Titanium Dioxide (TiO_2)-"

In handling nanofillers, it is very important to use their positive aspects as well as to ensure safety in the research stage and in practical use. Table 9.1 shows reports and guidelines regarding the handling of nanomaterials involving nanofillers in Japan. Moreover, Table 9.2 shows similar reports and guidelines published in other countries. An understanding of how to protect the human body and the environment is essential to accelerate the nanocomposite research and to expand practical applications. The experiment and manufacturing environment based on the reports and guidelines is important for safely handling nanofillers.

Table 9.2 Reports and guidelines regarding nanomaterial handling overseas

Organization/institute	Published year	Report/guideline
ISO (International Organization for Standardization)	2008 (TR 12885)	Nanotechnologies—Health and Safety Practices in Occupational Settings Relevant to Nanotechnologies
European Commission	2008	Regulatory Aspects of Nanomaterials
OECD (Organization for Economic Co-operation and Development)	2009	Comparison of Guidance on Selection of Skin Protective Equipment and Respirators for Use in the Workplace: Manufactured Nanomaterials
NIOSH in USA (National Institute of Occupational Safety and Health)	2009	Approaches to Safe Nanotechnology—Managing the Health and Safety Concerns Associated with Engineered Nanomaterials
BAuA in Germany (Federal Institute for Occupational Safety and Health)	2007	Guidance for Handling and Use of Nanomaterials at the Workplace
KOSHA (Korea Occupational Safety & Health Agency)	2007	Nanomaterial Hazardous Assessment and Prevention Strategy of Occupational Health Effect

9.2 Future Prospects

9.2.1 Worldwide Interest Continues to Be Enhanced Year by Year

T. J. Lewis published a paper entitled "Nanometric Dielectrics" in *IEEE Transactions on Dielectrics and Electrical Insulation* (hereinafter called *IEEE Trans. DEI*) in 1994 [7]. This paper proposed the concept of nanocomposite insulation materials and is considered to be their origin. After the publication of this paper, many studies on nanocomposite insulation materials have been conducted, and the number of related papers was 900 by 2014. Figure 9.6 shows the transition of the number of related papers from 1994 to 2014 [8]. (The figure contains additional data in 2014.) Papers were counted in *IEEE Trans. DEI* and proceedings of typical international conferences concerning dielectric and electrical insulation such as CEIDP, ISEI, ICSD, ICPADM, and ISEIM.

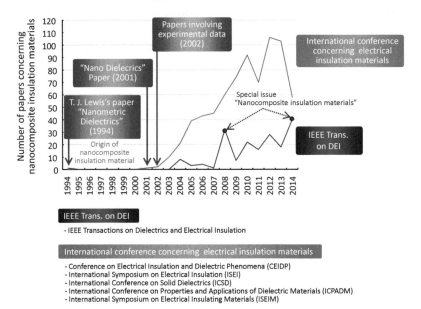

Figure 9.6 Transition of the number of papers concerning nanocomposite insulation materials. © 2013 IEEE. Reprinted, with permission, from Tanaka, T., and Imai, T. (2013). Advances in Nanodielectric Materials Over the Past 50 Years, *IEEE Electrical Insulation Mag.*, **29**(1), pp. 10–23.

After the publication of Lewis' paper in 1994, no paper was published until 2000. However, a paper that proposed the positive potential of nanocomposite insulation materials was published in 2001 [9], and two papers that showed experimental data on their insulation properties were published in 2002 [10–11]. Preparation and evaluation of many kinds of nanocomposites increased the number of related papers from 2002 to 2005. This tendency stimulated the activity in nanocomposite research, and the number of related papers rapidly increased after 2006. Moreover, a special issue on nanocomposite insulation materials was published in *IEEE Trans. DEI* in 2008.

After 2008, the number of related papers increased. However, the number of related papers in both *IEEE Trans. DEI* and the proceedings of typical international conferences briefly decreased in 2011. The number of related papers increased again in 2012, and the number in typical international conferences reached a record of 106. Recently, novel fabrication methods for nanocomposite insulation materials and measurements of their materials, computer simulation concerning the interfacial region between nanofiller and polymer and search for applications in industries have been studied. The fad of nanotechnology in 2000 has passed. However, worldwide interest in nanocomposite insulation materials remains a priority.

As mentioned above, nanocomposite insulation materials fascinate many researchers and engineers in the dielectric and electrical insulation field around the world. However, the answers to the questions "whether the research on nanocomposite insulation materials has been successful" and "whether the technology of nanocomposite insulation materials is really useful" depend on their practical use and their contribution to the industry.

9.2.2 Various Applications Are Explored for Their Realization

Chapter 2 of this book, entitled "Potential Applications in Electric Power and Electronics Sectors," introduced application research on nanocomposite insulation materials toward specific apparatuses and products. DC power cables insulated with XLPE nanocomposite and enameled wire coated with polyesterimide nanocomposite are already in use. However, some nanocomposites remain in the research stage. Figure 9.7 proposes future perspectives of nanocomposite insulation materials in power and industrial sectors, although the application research is in various stages [8]. Nanocomposite insulation materials will be a key technology to enable power apparatuses, cables, and motors to have superior performance such as high-voltage, large-capacity, high-efficiency, and long-life operation.

As mentioned in Chapter 1, energy consumption and oil demand have increased due to population expansion, giving rise to the serious worldwide problem of global warming. Global population will reach 9 billion in 2050, and energy consumption will increase further, as shown in Fig. 9.8 [12]. In general, it is said that there will be three problems in the twenty-first century: "economy," "energy," and "environment." These three "E" problems interact with each other as shown in Fig. 9.9. Thus, it is called the "trilemma of 3E." (Dilemma means a contradiction between two matters, but trilemma means a contradiction between three matters.) Cooperation on the "3E" is essential to realize a sustainable society.

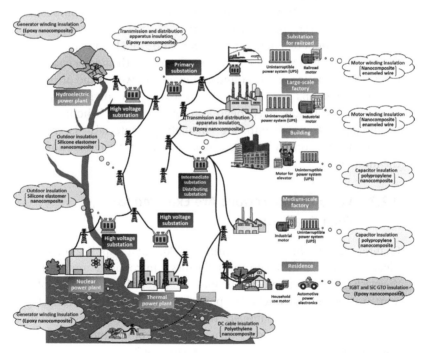

Figure 9.7 Future perspective of nanocomposite insulation materials in power and industrial sectors. © 2013 IEEE. Reprinted, with permission, from Tanaka, T., and Imai, T. (2013). Advances in Nanodielectric Materials Over the Past 50 Years, *IEEE Electrical Insulation Mag.*, **29**(1), pp. 10–23.

High-voltage, large-capacity, high-efficiency, down-sizing and long-life operation of power apparatuses, cables and motors thanks to nanocomposite insulation materials are expected to contribute to both energy saving and environmental harmony. When the future perspective of nanocomposite insulation materials in power and industrial sectors comes true in cooperation on the "3E," we will be sure to answer with confidence the above questions of "whether the research on nanocomposite insulation materials has been successful" and "whether the technology of nanocomposite insulation materials is really useful."

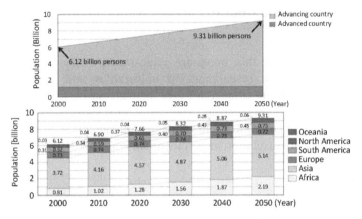

(a) Estimate of population increasing in world.

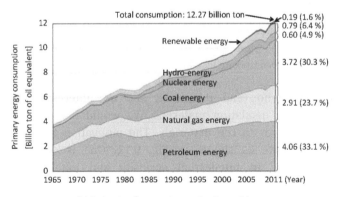

(b) Estimate of energy increasing in world.

Figure 9.8 Increase in energy consumption due to population expansion [12].

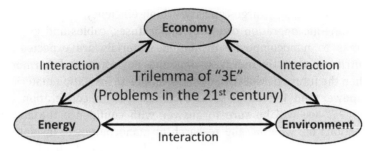

Figure 9.9 Trilemma of "3E" (three problems in the twenty-first century).

9.2.3 Open the Door to Polymer Nanocomposites for Future

Target readers of this book are educators in universities and technical colleges, researchers in universities and companies, students of graduate schools, and general readers who are interested in science and engineering. Therefore, this introductory book covers the range from applications to fundamentals concerning nanocomposite insulation materials, as shown in Fig. 9.10.

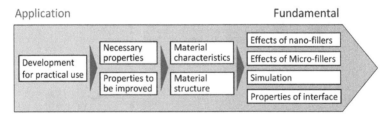

Figure 9.10 Flow from application to fundamental in research of nanocomposite insulation materials.

If those who read this book and are interested in the nanocomposite insulation materials, we would like to ask you to open the door to the materials. The motivation for research has two aspects as shown in Fig. 9.11. One is spontaneous motivation due to intellectual curiosity and inquiring mind. Curiosity and inquiring minds to clarify natural providence are driving forces toward the progress of science and technology and create today's society with comfort and convenience. In fact, curiosity and inquiring minds are important engines for research. Another is extrinsic motivation from the demands of society (problems in society). As mentioned above, nanocomposite insulation materials will contribute to energy and the environment to realize a sustainable society.

Figure 9.11 Motivations of research and development.

Both motivations continue with research and development of nanocomposite insulation materials too. Past research and developments have specified priority subjects, as shown in Fig. 9.12. Interdisciplinary research between electricity, chemistry, materials, physics, biology, and computer simulation and collaboration between the industry and academia are essential to tackle these subjects. Moreover, Chapter 3, entitled "Compatibility of Dielectric Properties with Other Engineering Performances," of this book describes new insulation materials with more than two functions by using nanofiller dispersion. Research on these multifunctional super-composite insulation materials has already started. This new insulation material field is young, deep, and large. Research on the range from fundamentals to application is necessary in this field. The computer simulation described in Chapter 8, entitled "Computer Simulation Methods to Visualize Nanofillers in Polymers," promises to act as an intermediary. Long-term research by new young researchers and engineers will be necessary in this field [13].

Two decades have passed since Lewis' paper in 1994, and more than a decade has passed since the paper that provided experimental data on nanocomposite insulation materials in 2002. The technology of nanocomposite insulation materials has shifted from "infancy" to the "establishing and developing period." It is essential to monitor the progress of this technology. We hope this book will become a milestone in the development of nanocomposite insulation materials.

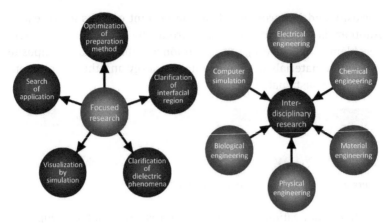

Figure 9.12 Priority subjects in research of nanocomposite insulation materials.

References

1. Ministry of Economy, Trade and Industry, Japan. (2007). *Guidebook on Chemical Risk Assessment for Business Operators* (in Japanese).
2. Kishitani, M. (2010). *Polyfile*, October, pp. 50–55 (in Japanese).
3. Chemical Management Center, National Institute of Technology and Evaluation, Japan. (2014). *Risk Assessment on Chemicals–For Better Understanding–*, September 2014 revised, p. 2.
4. Nakanishi, J. (2006). *Chemistry and Chemical Industry*, **59**, April, pp. 454–455 (in Japanese).
5. National Institute of Advanced Industrial Science and Technology (AIST), Japan. (2011). *Risk Assessment of Manufactured Nanomaterials*.
6. Kobayashi, T. (2007). *Expected Materials for the Future*, **7**(7), pp. 55–57 (in Japanese).
7. Lewis, T. J. (1994). Nanometric Dielectrics, *IEEE Trans. Dielectr. Electr. Insul.*, **1**(5), pp. 812–825.
8. Tanaka, T., and Imai, T. (2013). Advances in Nanodielectric Materials Over the Past 50 Years, *IEEE Electrical Insulation Mag.*, **29**(1), pp. 10–23.
9. Frechette, M. F., Trudeau, M., Alamdari, H. D., and Boily, S. (2001). Introductory Remarks on Nano Dielectrics, *Annual Rept. IEEE CEIDP*, pp. 92–99.

10. Imai, T., Hirano, Y., Hirai, H., Kojima, S., and Shimizu, T. (2002). Preparation and Properties of Epoxy-organically Modified Layered Silicate Nano Composites, *Proc. IEEE ISEI*, pp. 379–383.

11. Nelson, J. K., Fothergill, J. C., Dissado, L. A., and Peasgood, W. (2002). Toward an Understanding of Nanometric Dielectrics, *Annual Rept. IEEE CEIDP*, pp. 295–298.

12. The Federation of Electric Power Companies of Japan (2011). *Nuclear and energy drawings, Information Library*, p. 1-1-2, p. 1-1-7 (in Japanese).

13. Tanaka, T. (2009). Invitation to the Study of Advanced Polymer Ceramic Composite Dielectrics, *Proceedings of the 40th Symposium on Electrical and Electronic Insulating Materials and Application in Systems*, pp. 33–38 (in Japanese).

Index

alkylammonium ion, quaternary 160–161

alkylammonium ions 160, 164, 302–303

alumina 6, 10, 66, 89, 101–102, 155, 177, 185, 194–195, 236, 249, 299, 324

alumina fillers 14, 79, 81, 97, 300

alumina nanofiller 59, 194–195, 202, 240, 334

alumina nanofiller agglomerates 193

alumina nanoparticles 136, 177–178

alumina particles 177

aluminum 4, 9, 88, 101–102, 105, 131

aluminum nitride 28, 89, 92, 194–195

amine
  primary 290–291
  tertiary 290–291

antioxidant 118–119

arc, dry-band 58–59

arc discharge 54, 56–58, 63

ATH, micro-sized 266

barium titanate 78, 133, 135, 150, 304

base epoxy resins 9, 95, 97, 112, 237–238, 240, 290, 303, 305, 307, 310

base polymer 151, 217, 220, 222, 248, 259–261, 294, 324, 337

BN, *see* boron nitride

BN fillers 95, 98

boron nitride (BN) 10, 89, 91, 95, 98, 150, 175, 229

breakdown characteristics 224–226

breakdown strength 7–9, 31, 33, 94, 97–98, 230–231

breakdown voltage 82, 227, 229–230, 238

cables 2, 19, 21, 23, 25, 27, 29, 31, 33–34, 233, 245, 254, 304, 373, 375

calcium carbonate 292–293, 299

capacitance 130, 136, 184, 187

capacitor models 189–190

capacitors 30, 78, 136, 184, 187–189, 211

carbon 15, 233, 298–299, 318, 333

carbon black (CB) 298–299, 318–319

carbon fiber–reinforced plastics (CFRPs) 298–299

carbon nano-balls (CNBs) 176

carbon nanotube (CNT) 175, 318–319, 369–370

cation exchange capacity 159–160, 164

cation exchange capacity (CEC)
159–160, 164
CB, *see* carbon black
CEC, *see* cation exchange
capacity
ceramics 79, 102, 105, 111,
313
CFRPs, *see* carbon
fiber–reinforced
plastics
charged particles, collision of 49
Charpy impact strength 301,
309–310
of epoxy resin nanocomposite
309–310
chemical bonding, state of
341–342, 348
clay nanocomposites 158,
160, 163–165, 167, 169,
300
clays 4, 46, 73–74, 103, 107,
158–169, 172, 203, 291,
298–300, 302, 306, 311,
314, 316, 334
compound ornamentation 103
exfoliated 162–163, 166–167
exfoliation of 165–166
swelled 166–167
CNBs, *see* carbon nano-balls
CNT, *see* carbon nanotube
coarse-grained molecular
dynamics 352–361, 363
colloid science 1, 12–13
composite materials 2, 13–14,
72, 74, 87–89, 91, 93, 95,
97, 176, 206, 217, 298–299,
324
composites 13, 15, 77–78,
81, 88, 90, 93–95, 98,
100–101, 103, 105, 107,
110–111, 113, 115, 129,

131, 133, 135–136, 189,
192–193, 207, 229, 251,
298–300
magneto-dielectric 110–111
thermal conductivity of 90, 93
conduction current 210–211,
213, 215, 217, 219, 221
conductivity 113–114, 189,
199, 201–205, 207, 210,
255, 280–281, 328, 333
electrical 2, 110, 113, 115,
199, 202, 204–205, 330,
346
conductors 26, 29–30, 41, 49,
68–69, 100, 115, 206, 221,
274, 276–277
copper 76, 80, 88, 101, 131,
279–282
copper foil electrode 278–279
crack propagation 306–307,
318–320
cracks 46, 108, 306–307,
318–319
cross-linked polyethylene
(XLPE) 5, 30–31, 118,
151, 205, 215–216,
218–219, 231, 245, 256,
259–261, 350
crossover phenomenon
240–241
curing 75–76, 81, 122,
167–168, 288, 329
curing reaction 81, 167

DC power transmission 30–31
DC stress 214–218, 220–222
DC transmission 29–30
DC voltage 33, 186, 199, 227,
229, 273, 276–277
DC-XLPE 31–33

defect states, unoccupied interfacial 349–350
density functional theory (DFT) 346, 348, 359
DFT, *see* density functional theory
dielectric, composite 196–197
dielectric breakdown 95, 233, 240–241, 346
long-term 233, 235, 237, 239
dielectric breakdown strength 7, 94–95, 98, 245
dielectric breakdown time 95, 241
dielectric breakdown voltage 94, 96
dielectric constant 41, 44, 78–79, 83–84, 129–132, 134–136, 225, 304, 348–349
dielectric loss 4, 29–30, 183–187, 189, 191–193, 195
dielectric loss of nanocomposites 192–193
dielectric loss tangent 78, 83
dielectric materials 4, 211, 346
polymer/nanocomposite 204
dielectric nanocomposites 308
dielectric permittivity 110–111
dielectric polarization 184–186
dielectric properties 184, 187, 189, 191–192, 194, 199, 329–330
dielectric spectra 187
dielectric spectroscopy 183, 330
dielectric strength 5
differential scanning calorimetry (DSC) 286–287, 289
dissipative particle dynamics (DPD) 356

DPD, *see* dissipative particle dynamics
DSC, *see* differential scanning calorimetry

electric breakdown 210, 215–216
electric charge 43–44, 280
electric charge distribution 279
electric discharge 263–264, 269
electric field
direction of 185–187
enhanced 215, 219
local 215, 217, 219
electric field distribution 7, 32–33, 210, 212, 214, 220
electric field enhancement 218–219
electric field strength 44, 272, 277
electric pulses 244–245
electric stress 6, 10, 213, 235, 238, 250
electric stress resistance 100, 102–103
electric vehicles 38
electrical breakdown 213–214, 224–226
electrical breakdown strength 204, 226, 229, 231
electrical insulation 1, 7, 54, 336, 372
electrical resistivity 147, 207
electrical signals 129
electrical stress resistance 103, 105
electrochemical migration 70, 272–273, 275, 277, 279, 281

384 | Index

electrochemical migration resistance 68
electrode systems 95, 226, 233–234, 245–247
  parallel-plate 228–229
electrodes
  circular metal 187–188
  comb-form 277
  needle-plate 106
electromagnetic waves 39, 187
electron avalanche 238–239
electron beams 333, 335
electron injection 205, 234
electron microscopy 333–335
electron trap sites 349
electronic devices 68–69, 71, 73, 75, 77, 79, 81, 83, 115, 244, 287, 299, 311
electronic equipment 79, 272–273, 277
electronic polarization 185–186
electronics devices 68–70, 72, 84
electrons 171, 226, 238, 241, 247, 273, 330, 333–334, 349, 358–359
  injected 235
enamel 42, 44
enameled wire 27, 38–40, 42–43, 46–47, 49, 154, 373
energy consumption 319, 374–375
EOT, see equivalent oxide thickness
epoxy 13, 26, 73, 75–76, 88, 90–93, 101–102, 120, 136, 147, 151, 154–155, 162, 166, 236, 247–249, 251, 291, 305–306, 309, 330, 342
  base 237, 240, 248, 330
epoxy/alumina nanocomposites 120

epoxy-based nanocomposites 162–163, 165, 308
epoxy nanocomposites 3, 72, 237, 239, 288, 292
epoxy resin 3, 5, 8–10, 20, 25, 54, 56, 71–76, 78, 83, 92–95, 97–98, 101–103, 105–106, 112, 114, 120–121, 124–127, 136, 148, 151, 162, 166–167, 172, 184–185, 190–191, 193–195, 229, 234, 238, 247, 266, 278, 280, 282, 286–288, 291–292, 295, 303, 306, 317–319, 329, 334–335, 338, 341, 360–361
  alkoxysilane-bonded hybrid 121
  alumina-filled 94
  cast 24
epoxy resin nanocomposites 121, 124, 302–303, 309–310
epoxy resin/ZnO nanocomposite 341–342
epoxy/silica nanocomposite 152–153, 248, 280, 308
equivalent oxide thickness (EOT) 131
erosion 40, 45, 49, 54, 56–57, 63, 95, 237, 240, 263–266, 269
erosion degradations 263–264, 268–269
ethylene-vinyl acetate copolymer 54, 122
exfoliation 25, 163–167
exfoliation methods 163–164

fatigue 302, 315, 318

fatigue properties 301, 315–318, 320

fatty acids 292–293

Fe nanoparticles 113, 115

Fe particles 112

FEM, *see* finite element method

filler permittivity 220

fillers
ceramic 102, 188
conductive 206
granular 150
high-dielectric-constant 132
inorganic nano-size 217, 219
insulating 206
intercalated flat 46–47
microsilica 335
nano-sized 34, 119, 354–355

finite element method (FEM) 133–134, 347

fracture surfaces 307, 318

fracture toughness 298, 301–302, 306–308

fractures 300, 302, 315

gas-insulated switchgear (GIS) 101

gas insulation systems 20

gaseous discharges 244, 246, 251

gate insulator film 130–131

GEP, *see* global warming potential

GIS, *see* gas-insulated switchgear

global warming potential (GEP) 20–22

greenhouse gas 21–22

guanidine 73–74

HDT, *see* heat distortion temperature

heat distortion temperature (HDT) 165

heat resistance 2, 14–15, 68, 70–72, 76, 78, 82–84, 119–122, 124–125, 127, 130, 263–264, 268–270, 286–287

Hf-based oxides 130–131

high-dielectric-constant capacitor materials 78

high-dielectric-constant nanofiller 132

high dielectric permittivity 110–111, 113, 115

high heat-resistant composites 118–119, 121, 123, 125

high thermal conductivity 10, 15, 28, 68–69, 79–82, 88, 91, 95, 97

high-thermal-conductivity insulation material 28

high-thermal-conductivity polymer composites 91

high-voltage arcing 64–65

high voltage endurance 100, 103, 105, 107

humidity 41, 43–45, 200, 277
relative 44, 144, 330

hydrophilic nanosilica 258

impact-resistant polymer alloy material 299

impulse breakdown strength 8, 215, 217

inorganic fillers 2, 6, 11–13, 33–34, 71–72, 81, 264, 269, 325–326
conductive 222

inorganic materials 2, 6, 150, 250

high-dielectric 129
inorganic nanofillers 6, 10, 31,
    237, 264, 324
inorganic particles 45, 49, 173
insulated transformer 21
insulating materials 1, 7, 16, 44,
    98, 124, 184, 186, 189, 204,
    210, 212, 217–218, 222,
    247, 249, 264, 287, 311,
    333, 346
  solid 8, 190, 210
insulating polymers 101
  high voltage endurance of 101
insulation breakdown 27,
    39–40, 47, 50
insulation breakdown strength
    106, 230
insulation breakdown time 104,
    106–107
insulation degradation 39, 44,
    84, 244, 253, 263
insulation design 8, 39, 50
insulation deterioration 272,
    277
insulation materials 30, 100,
    285–287, 372
  conventional 28, 171
  nanocomposite 374–376, 378
  solid 31
  super-composite 377
insulators
  electrical 184, 189
  outdoor 263, 265
  polymeric hollow-core 263
  porcelain 54–55
interactions, polymer-nanofiller
    327
interatomic distance 102, 353
interfaces
  electrode/material 220
  nanofiller/polymer 346

interfacial interactions 295–296,
    326–327
interfacial model, two-layered
    327
interfacial polarization 185–186,
    188–189, 348
interfacial structures,
    polymer/nanofiller 15
isothermal thermogravimetric
    analysis (ITA) 293–294
ITA, see isothermal
    thermogravimetric
    analysis

laser irradiation 58, 61
layer-structured fillers 158, 162
layered silicate (LS) 4, 6, 9, 46,
    48–50, 106, 108, 123–124,
    236, 266, 293, 324,
    334–335, 339
layered silicate clays 72–73
LDPE, see low-density
    polyethylene
LDPE/MgO Nanocomposite 153,
    217, 220, 222
low-density polyethylene (LDPE)
    3, 7–8, 111, 153, 192–193,
    205, 213–218, 220,
    256–258, 260–261
low electric field conditions
    240–241
low electric field conduction
    199, 201, 203, 205
LS, see layered silicate

magnesia 6–7, 185, 192,
    194–195, 256, 324
magnetic materials 110–111

material degradation 71, 235,
    244–247, 249, 253, 255,
    257, 259, 263, 265, 267,
    269, 272–273, 275, 277,
    279, 281
materials
  bulk 111, 150, 347
  electronic 127, 142, 147–148
  ferromagnetic 111
  high-dielectric-constant 129
  low-dielectric-constant 129,
    131
  low loss magneto-dielectric
    117
  metallic 315, 318
  organic 2, 5, 45, 49, 56
  polymer alloy 299–300
matrix polymer chains 12
MD, see molecular dynamics
metal alkoxides 3, 142
metal conductors 9, 24–25
microcomposites 96–97,
    152–153, 188, 191–192,
    194–195, 230, 269, 307,
    310–311, 330
microfillers 10, 26, 57–62, 66,
    71, 84, 88, 94–95, 97–98,
    100, 105, 108, 193, 229,
    285, 288, 292, 309, 324,
    366
  inorganic 25, 264
  magnesia 192
microparticles 204, 335
MO, see molecular orbital
MO method 345, 347, 349–351,
    355
molecular dynamics (MD)
    346–347, 351–361, 363
molecular dynamics simulation
    352, 354
molecular orbital (MO) 346–347,
    353

nano-sized silica 267, 269
nanoclay 190–191, 291
  dispersion of 290
nanocomposite dielectrics 15
  fabricated 15
nanocomposite-enameled wires
    45–47
nanocomposite functionalization
    199, 204–205
nanocomposite insulation
    materials 336, 365–366,
    372–378
nanocomposite insulators 63,
    224, 226–228
nanocomposite materials
  epoxy/silica 78
  innovative multifunctional
    polymer 351
  magneto-dielectric 111
nanocomposite resin 355, 359
nanocomposite surge-resistant
    wires 50–51
nanocomposite wires 42, 44–45,
    47–50
nanocomposites 127, 165, 258
  cured 337–338
  magneto-dielectric 110, 112
  micro-silica-filled 9
  polyamide/clay 14, 169,
    299–300
nanofiller addition effects 250
nanofiller agglomerates 194,
    334
nanofiller aggregates 229, 237
nanofiller dispersibility 60–61
nanofiller dispersion 25, 54, 57,
    59–61, 63, 120, 287–288,
    290, 292, 296, 302, 305
nanofiller handling 365, 367,
    370–371
nanofiller interfaces 184, 191,
    196, 258–259, 261, 341

nanofillers 173, 175, 177
  hydrophobic 259
  magnesia 192, 256–257, 260
  quasi-spherical 150–151, 153
  surface modification of 171,
    173, 175, 177
nanohybrid resin 125–126
nanomaterials, industrial 369
nanomicrocomposites 9, 25–27,
    70, 80, 84, 95–97, 290–291,
    293, 310, 335
  silicone 292
nanoparticles, inorganic 45, 49
nanosilica 9–10, 107, 120–121,
    267
  hydrophobic 258
NOAEL, *see* no observed adverse
    effect level
nylon 123, 298, 316–317
no observed adverse effect level
    (NOAEL) 367–368

octadecylamine 203, 291
Ohm's law 210, 212–213
oleylamine 112–113, 115
organic insulating materials,
    resistance enhancement
    of 169
organic modifier 103, 160–161,
    164
organic polymers 10, 269,
    324–325, 354
oxygen 235, 245, 247, 314, 359

parallel-plane electrodes
    199–200
partial discharge 25, 38–41, 45,
    49–51, 169, 244, 339

partial discharge
    characteristics 43–44
partial discharge inception
    voltage (PDIV) 39, 41–45
partial discharges (PDs) 6, 234,
    237–238, 241, 244–247,
    249–250
PDIV, *see* partial discharge
    inception voltage
PDs, *see* partial discharges
PEA, *see* pulsed electro-acoustic
plasma 176, 339
plastics 298, 301
polarization 185–187, 191, 196
  atomic 185–186
  dipole 185–187, 194
  space-charge 185–186
polyamide 4, 133, 151, 162,
    165, 169, 205, 298, 300,
    330, 339
polyamide-based clay
    nanocomposite 162–163,
    165
polyamide nanocomposite 309
polyethylene 6, 88, 118, 151,
    185, 191–192, 217, 229,
    234–235, 238, 247–248,
    253, 260, 294, 329
polyimide 45, 78, 135, 146–147,
    151, 202, 205, 350
polymer alloys 14, 299
polymer base resins 7
polymer chains 10–11, 234,
    260–261, 289, 296, 323,
    326, 329
polymer composites 87, 129,
    298–299
polymer insulator, structure
    of 55
polymer insulators 5, 54–57,
    66, 68, 254, 263, 265

polymer insulators for outdoor
use 54–55, 57, 59, 61, 63,
65
polymer materials 55–57
organic 245
polymer matrices 11, 63,
91–92, 204, 227,
327–330, 366
polymer nanocomposite
insulation 5
polymer nanocomposite
technology 313
polymer nanocomposites
application of 5
characteristics of 6
commercialized 147–148
cured 144
macroscopic performance
of 325, 328
silicone-based 56
polymer/nanofiller
interfaces 323–324, 326,
328, 330, 334, 336, 338,
340, 342
polymer nanomaterials 120
polymer nanomicrocomposites
69
polymer resins 6, 9–10, 81
pure 1
polymer thermal characteristics
286–287, 295–296
polymerization 123, 162–165
polymers
acrylic 75–76, 148
dielectric properties of 191,
196
electrical insulating 111
matrix 147, 162
non-polar 145, 350
silicone 263, 266
power apparatuses 9, 23–25,
100, 102, 233, 373, 375
electric 244, 311, 313

power electronics 70–71
preparation of polymer
nanocomposites 141–142,
144, 146, 148, 150, 152,
154, 156, 158, 160, 162,
164, 166, 168, 172, 174,
176, 178
printed wiring boards (PWB) 5,
10, 91, 95, 245, 272–274,
277–278
pulsed electro-acoustic (PEA) 7,
214, 278
PWB, *see* printed wiring
boards

quantum mechanics 323,
328–329, 345, 347, 358

relative permittivity 114,
130–131, 133–135, 138,
184–186, 188–197
relative permittivity of
nanocomposites 191–195
resin 6, 9, 26, 68–69, 71, 73,
75–76, 79–81, 97, 102, 107,
113, 120, 125, 132, 151,
162, 171–173, 192,
250–251, 257, 286, 288,
304, 318, 336, 340,
354–355, 359, 361–363
composite 73, 79
phenolic 13, 122, 125, 278
polymer insulator silicone 6
resin/filler interface affinity 304,
306
resin materials 77, 81, 127
resin substrates 78–79
room temperature vulcanization
(RTV) 57, 266

R

RTV, *see* room temperature vulcanization
RTV silicone rubber 266–267
rubber 65, 317–318

S

scanning electron microscopy (SEM) 26, 63, 113, 151, 176, 254, 280–281, 307, 318, 333–334, 336, 352
sealing resins 70–72, 300
SEM, *see* scanning electron microscopy
semiconductors 75–76, 84, 129, 131, 300
SF_6 20–22
SF_6 gas 20, 22–24, 28, 101
SF_6 gas insulation systems 23, 25
short circuits 272, 276–277
short-term breakdown characteristics 224–227, 229, 231
short-term breakdown characteristics of nanocomposite insulators 226, 228
silane 8, 91–93, 125, 173, 325
silane coupling 94, 172, 249, 325
silane treatment 172, 307
silica 6, 9, 14, 42–43, 45, 47, 72, 89, 101–102, 121, 129, 142, 148, 172–173, 175, 185, 194–195, 256, 266–268, 270, 291, 299–300, 306, 324, 349
 amorphous 46, 48–50
 colloidal 153–154, 340
 fumed 59, 120, 151
 micro-sized 266–267
 nano-fumed 292–293
 natural 59

silica fillers 57, 71, 78, 105, 153
silica microfiller 290–291, 307, 311
silica nanofiller 61, 66, 259, 280, 282, 289, 292, 295, 307, 338, 359–361
 fumed 58, 64
 hydrophobic 360–361
 surface-treated 294
 surface-untreated 294
silica nanoparticle 172, 174, 331
silicon rubber 61–62
silicone 64, 71–72, 293
silicone gels 71–72, 75
silicone nanocomposite insulation 59–62
silicone nanocomposites 54
silicone resin 119–120
silicone rubber 54, 56–64, 88, 120, 151, 263, 265–266, 269–270
silicone rubber nanocomposites 57–60, 63–66
 thermal degradation properties of 64–65
silicone/SiO_2 nanocomposites 64–65
SiO_2 3, 6, 25, 43, 111, 118, 130, 142, 146, 150, 159, 175, 195, 230, 294, 348–349
SiO_2 microfillers 24–26
SiO_2 nanofillers 146
SIS, *see* solid insulated switchgear
sodium ion 159–160, 162
sol-gel method 3, 6, 141–143, 145, 147–148
solid insulated substation 28–29
solid insulated switchgear (SIS) 20, 24–26
solid insulation 7, 200, 244
space charge 6–7, 32–33, 43, 49, 210, 212–213, 215, 217, 278–279

space charge accumulation 20, 31, 33, 210–211, 213–219, 221

space charge accumulation behaviors 216, 218–219

surface wettability 44

surge, inverter 38–40, 51, 245

surge-resistant wires 45, 47

switchgear, insulated 21, 23–24, 28

switchgear insulation 5

TEM, *see* transmission electron microscopy

tensile modulus 77, 302–303

TEOS, *see* tetraethoxysilane

TEOS-PDMS nanocomposites 145–147

tetraethoxysilane (TEOS) 3–4, 142–143, 145, 147

TGA, *see* thermo gravimetric analysis

thermal behavior 286–287

thermal characteristics 285–287, 292, 296

thermal conductivity 5, 72, 81–84, 87–93, 95, 98, 175, 178

thermal decomposition 64–65, 121, 247

thermal degradation 59–60, 63–64, 119

thermal degradation temperature 286–287, 295–296

thermal diffusivity 88–89

thermal endurance 120, 293, 313–314

thermal expansion 25, 73, 100–101, 105, 108, 286

linear coefficient of 102, 105–106

thermal expansion coefficient 9, 70, 78, 83–84, 100–103, 105, 108, 233

thermal performance 285, 287, 289, 291, 293, 295

thermo gravimetric analysis (TGA) 59, 121, 160–161, 269, 286, 293

thermoplastic resins 150–151, 153, 317

thermosetting resins 71, 102, 125, 150–151, 286, 354–355

TiO_2 3, 6, 43, 111, 132, 150, 369–370

titania 6, 8, 42–43, 47, 66, 150, 185, 249, 324, 369

transformers 20–21, 24, 28

oil-filled 28

transmission electron microscopy (TEM) 26, 113, 153, 172, 334, 336–337

tree channels 237–239, 241–242

tree initiation 7, 233–236, 238, 241, 330

tree initiation time 238, 241

tree propagation 233, 236, 240–242

treeing 7, 26, 233, 241

electrical 25, 70, 107

treeing breakdown 233–234, 236

trees

bow-tie 254–256

electrical 26, 230, 233, 253, 259

initial 238–239

minute 239, 241

Triton addition 61–62

Triton concentration 61–62
Tungsten electrode 264–265

ultrasonic waves 154–156
ultraviolet radiation 54, 56

vacuum interrupters 24–26
vapor phase epitaxy carbon fiber (VGCF) 318–319
VGCF, *see* vapor phase epitaxy carbon fiber
voltage source converter (VSC) 31
VSC, *see* voltage source converter

water 142–143, 176, 193, 247, 253–255, 258–259, 264, 273, 275–276, 330, 337, 366
water absorption 194, 231, 258–259

water electrodes 255–256
water trees 174, 253–261
  multiple 255–256
  single 255–256
Weibull plots of insulation breakdown strength 106, 230
wires
  conventional 45, 48–49
  general-purpose 42, 44, 47–48

X-ray diffraction (XRD) 160–161, 339
XLPE, *see* cross-linked polyethylene
XLPE cables 245, 253–254
XLPE nanocomposites 31, 373
XRD, *see* X-ray diffraction

Young's moduli 175, 301, 304–305, 355